# Service-Learning in Undergraduate Geosciences



Alexandra Beatty, *Rapporteur*

Board on Science Education

Division of Behavioral and Social Sciences and Education

*The National Academies of*
SCIENCES · ENGINEERING · MEDICINE

THE NATIONAL ACADEMIES PRESS
*Washington, DC*
**www.nap.edu**

**THE NATIONAL ACADEMIES PRESS 500 Fifth Street, NW Washington, DC 20001**

This activity was supported by Grant No. 1544268 from the National Science Foundation. Any opinions, findings, conclusions, or recommendations expressed in this publication do not necessarily reflect the views of any organization or agency that provided support for the project.

International Standard Book Number-13: 978-0-309-45272-4
International Standard Book Number-10: 0-309-45272-4
Digital Object Identifier: 10.17226/24621

Additional copies of this publication are available for sale from the National Academies Press, 500 Fifth Street, NW, Keck 360, Washington, DC 20001; (800) 624-6242 or (202) 334-3313; http://www.nap.edu.



Printed in the United States of America

Suggested citation: National Academies of Sciences, Engineering, and Medicine. (2017). *Service-Learning in Undergraduate Geosciences: Proceedings of a Workshop.* Washington, DC: The National Academies Press. doi: 10.17226/24621.

*The National Academies of*

# SCIENCES • ENGINEERING • MEDICINE

The **National Academy of Sciences** was established in 1863 by an Act of Congress, signed by President Lincoln, as a private, nongovernmental institution to advise the nation on issues related to science and technology. Members are elected by their peers for outstanding contributions to research. Dr. Marcia McNutt is president.

The **National Academy of Engineering** was established in 1964 under the charter of the National Academy of Sciences to bring the practices of engineering to advising the nation. Members are elected by their peers for extraordinary contributions to engineering. Dr. C. D. Mote, Jr., is president.

The **National Academy of Medicine** (formerly the Institute of Medicine) was established in 1970 under the charter of the National Academy of Sciences to advise the nation on medical and health issues. Members are elected by their peers for distinguished contributions to medicine and health. Dr. Victor J. Dzau is president.

The three Academies work together as the **National Academies of Sciences, Engineering, and Medicine** to provide independent, objective analysis and advice to the nation and conduct other activities to solve complex problems and inform public policy decisions. The National Academies also encourage education and research, recognize outstanding contributions to knowledge, and increase public understanding in matters of science, engineering, and medicine.

Learn more about the National Academies of Sciences, Engineering, and Medicine at **www.national-academies.org**.

*The National Academies of*

SCIENCES · ENGINEERING · MEDICINE

## STEERING COMMITTEE ON SERVICE-LEARNING IN UNDERGRADUATE GEOSCIENCES—A WORKSHOP

**CATHRYN MANDUCA** (*Chair*), Science Education Resource Center, Carleton College
**ANTONY BERTHELOTE,** Department of Hydrology, Salish Kootenai College
**SUE EBANKS,** Department of Marine and Environmental Sciences, Savannah State University
**SARAH FORTNER,** Department of Geology, Wittenberg University
**ART GOLDSTEIN,** College of Science and Mathematics, Bridgewater State University
**EDWARD LAINE,** Department of Earth and Oceanographic Science, Bowdoin College
**JOHN SALTMARSH,** New England Resource Center for Higher Education, University of Massachusetts
**SUSAN SULLIVAN,** Cooperative Institute for Research in Environmental Sciences, University of Colorado

**KERRY BRENNER,** *Study Director*
**SAMMANTHA MAGSINO,** *Senior Program Officer,* Board on Earth Sciences and Resources
**HEIDI SCHWEINGRUBER,** *Director,* Board on Science Education
**MIRIAM SCHEIBER,** *Program Assistant*

## BOARD ON SCIENCE EDUCATION

# Acknowledgments

This proceedings of a workshop has been reviewed in draft form by individuals chosen for their diverse perspectives and technical expertise. The purpose of this independent review is to provide candid and critical comments that will assist the institution in making its published proceedings as sound as possible and to ensure that the proceedings meets institutional standards for clarity, objectivity, and responsiveness to the charge. The review comments and draft manuscript remain confidential to protect the integrity of the process. We wish to thank the National Academies Research Center for their work gathering literature related to service-learning for the committee's considerations during the workshop planning process.

We also wish to thank the following individuals for their review of this proceedings: Jonathan M. Harbor, Department of Earth, Atmospheric, and Planetary Sciences, Purdue University; Alexandra Moore, Department of Geological Sciences, Cornell University; Linda Silka, Senator George J. Mitchell Center for Sustainability Solutions, University of Maine; and Susan Sullivan, Cooperative Institute for Research in Environmental Sciences, University of Colorado Boulder.

Although the reviewers listed above have provided many constructive comments and suggestions, they did not see the final draft of this proceedings of a workshop before its release. The review of this proceedings was overseen by Julie C. Libarkin, Geocognition Research Laboratory and Department of Earth and Environmental Sciences, Michigan State

University. She was responsible for making certain that an independent examination of this proceedings was carried out in accordance with institutional procedures and that all review comments were carefully considered. Responsibility for the final content of this proceedings rests entirely with the rapporteur and the institution.

# Contents

**Appendixes**

# 1

# Introduction

The term "service-learning" generally refers to projects planned as components of academic coursework in which students use knowledge and skills taught in the course to address real needs in their communities. This kind of learning experience, which allows students to focus on critical, reflective thinking and civic responsibility, has become an increasingly popular component of undergraduate science education.

Service-learning projects may involve many sorts of activities that link students directly with people and groups in their communities so they can collaborate to solve problems. What is generally characteristic of service-learning projects is that they allow students to connect their coursework with real-world problems, to see how information they learn in class can be applied in solving problems and improving their communities, and to reflect on what they learn from the experience. For example, students might conduct a greenhouse gas emissions inventory for their college campus; investigate contaminant transport in a local watershed and identify potential risks to the local population; or examine the impact of geologic hazards on the human population and design appropriate responses.

Service-learning fits well with the goals for undergraduate study in the geosciences, the disciplines concerned with the Earth, which include geology; geophysics; geochemistry; and atmospheric, polar, and ocean studies.[1] Researchers in these disciplines explore natural processes and

[1]In this document, the term "geosciences" refers to the disciplines served by the Geosciences Directorate at the National Science Foundation, the sponsor of this project.

help people understand their relationship to the environment in which they live—they help people locate and use needed resources, prepare for natural disasters, and understand changes in climate, for example. Students in geoscience fields study issues, such as landscape evolution, environmental challenges, energy use and production, weather and climate, and their effects on society. These kinds of topics naturally highlight the direct links between academic studies and real-world issues.

Students doing service-learning projects as part of their geoscience education work with their faculty advisors to identify projects that address real-world issues, are of genuine importance in their communities, and draw on the knowledge and skills learned in geoscience courses. The students are given some autonomy to devise solutions to problems and are called on to reflect on how they have approached the task, the successes and challenges they encountered, and what they learned from the experience.

Geoscience educators have hoped that service-learning could help them to engage more undergraduate students, particularly women and members of underrepresented population groups, in these disciplines. However, the best ways to use service-learning as a recruiting tool and its potential impact are not known. There have been few systematic efforts to examine how service-learning is used in the geosciences and how it benefits students or to assess the lessons to be learned from similar programs in other disciplines.

With the support of the National Science Foundation, the National Academies of Sciences, Engineering, and Medicine planned a workshop to explore the current and potential role of service-learning in undergraduate geoscience education. The Committee on Service-Learning in the Undergraduate Geosciences was appointed to plan the April 2016 workshop. This committee included researchers and practitioners with expertise in undergraduate instruction in geoscience and other science disciplines, higher education administration, and service-learning in the geosciences and other fields. The committee was given the charge of exploring how service-learning is being used in geoscience education, its potential benefits, and the strength of the evidence base regarding the nature and benefits of these experiences; see Box 1-1 for the statement of task.

This proceedings document, prepared by the workshop rapporteur, summarizes the presentations and discussions that took place.[2] The planning committee's role was limited to planning and convening the

[2]The workshop agenda, biographical sketches of committee members and presenters, and a list of workshop participants can be found in Appendixes A and B. These and other resources, including the three papers commissioned for the workshop, can be found at the

> **BOX 1-1**
> **Statement of Task**
>
> An ad hoc committee will plan and conduct a 2-day public workshop that will explore how service-learning is being used in geoscience education, the potential benefits, and the strength of the evidence base. Presenters and participants at the workshop will discuss which student outcomes are most likely to be influenced by a service-learning experience, what "best practices" might look like, the potential challenges to implementing service-learning in undergraduate geoscience courses, and areas where research is needed. The workshop will feature invited presentations and discussion. It will provide an opportunity for discussion between expert researchers and practitioners in complementary fields, such as undergraduate geoscience, higher education administration, service-learning both within geoscience and in other fields, science learning at the undergraduate level, and higher education administration (to address infrastructure and resource issues).

workshop. The views contained here are those of individual workshop participants and do not necessarily represent the views of all workshop participants, the planning committee, or the National Academies of Science, Engineering, and Medicine.

The sponsor proposed seven sets of questions for the committee to consider in planning the workshop:

1. How does the geoscience community define and implement service-learning? What types of service-learning activities are currently being implemented within the geoscience community? How is service-learning being used in other science disciplines?
2. What are the learning goals of service-learning experiences and what student outcomes are measured? How consistent are the goals and outcomes across programs? What is the evidence that these goals and outcomes are being achieved?
3. What is known about best or effective practices in geoscience service-learning programming (e.g., identify design principles, describe what students benefit the most, and explore the influence of handbooks and toolkits)?
4. Is there evidence that these types of experiences lead to increased engagement/attainment in the geosciences? Is there evidence

project Website, along with an archived video of the proceedings; see http://sites.nationalacademies.org/DBASSE/BOSE/DBASSE_171478 [December 2016].

about the impact on attracting underrepresented minority students to geoscience fields? Is there evidence to indicate the point in a student's educational path at which these kinds of experiences have the most impact?

5. How does service-learning fit within the preparation of future geoscientists? Would such an activity be perceived positively by future employers?
6. What are the obstacles to implementing these programs and expanding access to successful programs or approaches to programing?
7. What research is needed to improve understanding of the potential contribution of service-learning in the geosciences?

The committee commissioned three papers to look in depth at aspects of this charge and structured the workshop presentations and discussions to explore these questions.[3] The three papers explored research and other insights from service-learning experiences outside of the geosciences (Silka, 2016), themes in the literature on service-learning within the geosciences (Savanick Hansen and Fortner, 2016), and what is known about the participation of historically underrepresented groups in the geosciences (O'Connell, 2016).

This proceedings document discusses the primary topics addressed at the workshop. Chapter 2 explores reasons why service-learning is important and ideas from other disciplines that may be useful for geoscience educators. Chapter 3 discusses what is known about two goals for geoscience service-learning programs: engaging diverse student populations and providing all students with positive outcomes. Chapter 4 focuses on the nature of these programs, using discussion of examples to flesh out the picture of how they are defined and implemented. The final chapter looks at the outlook for such programs in the future and questions for further research. Participants suggested numerous resources in the course of the workshop, and these are provided in Appendix C.

[3]The three commissioned papers and other resources can be found at the project Website, http://sites.nationalacademies.org/DBASSE/BOSE/DBASSE_171478 [August 2016].

# 2

# The Role of Service-Learning in the Geosciences

Service-learning has been a part of undergraduate teaching for decades, although it has not always been known by that name. With roots in education reforms of the 1960s and 1970s, the idea of formally incorporating service with pedagogical goals gained popularity during the 1990s.[1] The first workshop session provided an overview of the role this approach plays in undergraduate geoscience fields and lessons to be learned from its role in other fields of study. Two presenters explored the value and characteristics of service-learning in the geosciences: committee chair Cathy Manduca of Carleton College offered broad observations about key benefits of service-learning in the geosciences, and David Szymanski of Bentley University provided a more detailed look at how service-learning programs can work and the value they bring. A third presenter, Linda Silka of the University of Maine, drew both on research and her own experience with undergraduate service-learning in a variety of fields to identify themes that could be helpful in the geoscience context. The session closed with an opportunity for general discussion of strengths and weaknesses in the ways service-learning is currently done.

[1]For information about the history of service-learning, see a summary on the University of Southern California Website at https://dornsife.usc.edu/history-of-service-learning [October 2016].

## VALUE OF SERVICE-LEARNING IN THE GEOSCIENCES

Service-learning experiences may not only help students to integrate their academic learning with the world around them, Manduca observed, they also have benefits for institutions. They give colleges and universities opportunities to connect with the communities of which they are a part, and they can enrich geoscience departments by engaging students from diverse cultural backgrounds and academic disciplines. Manduca noted that the geosciences include not only study of the solid Earth, the atmosphere, and the oceans, but also study of the people who live on the Earth, which means study of geography, environmental sciences, and sustainability as well. The fields that constitute the geosciences, she commented, offer fertile ground for service-learning programs that address intersections between science and society.

Szymanski drew on experience with his own students at Bentley University to illustrate his thoughts about the value of service-learning in geoscience fields. He began by noting that service-learning has a more specific meaning than community service, with which it can be confused. Service-learning is linked with academic learning objectives, he explained, and helps students to develop professional skills—such as technological or scientific consulting, grant-writing, assessments, and research—that are not always aspects of community service experiences.

Students who participate in service-learning engage in the civic life of their communities in two ways, Szymanski suggested. They become involved in issues of communal importance (such as risks from natural hazards, waste disposal, conservation, and sustainability), and they may also participate in public decision making and the political process. It is through these experiences, he explained, that service-learning has the potential to teach students how to apply ideas from the disciplines they are studying, and perhaps also to develop the ability to synthesize ideas and skills from multiple disciplines in order to solve a problem. Students who engage in these projects learn about problems that are important to their community partners, Szymanski added, which encourages them to think deeply about different perspectives and how their own contributions may fit with work already being done to benefit a local community or perhaps beyond.

Szymanski noted that while the relevance of the geoscience disciplines to everyday life is not widely recognized, they are critical for understanding and addressing many societal and economic problems. Moreover, in his view, the geosciences "do a particularly good job with complexity—with wicked problems."[2] For example, he noted, many issues that come

---

[2]The term "wicked problems" originated in the study of management and refers to problems that are unusually challenging. Szymanski noted that even defining these problems

before elected officials or juries involve both political and legal challenges and also complex scientific understanding. Service-learning not only builds on the geoscience tradition of integrating ideas and skills from different disciplines, it also provides a public demonstration of the contributions the geosciences can make.

Szymanski noted that the challenges of environmental sustainability illustrate particularly well the important contribution of the geosciences. Increasingly, he noted, sustainability is recognized as a complex, or wicked, challenge that will affect virtually every kind of industry and business, as well as other aspects of society. Many business leaders are realizing that sustainability has come to be a factor in all of their decisions and that it is no longer adequate to address environmental concerns through tangential projects such as providing donations to an environmental cause or event. Businesses and other employers with this view recognize that they need workers in every kind of role who have the capacity to understand scientific issues and make sustainable decisions.

As more employers come to value this attribute in their employees, Szymanski continued, preparation for thinking about sustainability will be increasingly important for undergraduate students, regardless of their majors. To help prepare students, Bentley University—a business school—and some other institutions have focused on the goal of "reciprocal integration." This is an educational model in which business students spend time on courses in disciplines within the arts and sciences, and students from those disciplines take business courses, Szymanski explained. While the majority of Bentley's 4,250 students major in a business field, he noted, approximately one-third also complete coursework for a second concentration or major in the arts and sciences. He said that this approach develops four modes of thought:

1. practical reasoning, the application of technical training;
2. analytical thinking and problem solving;
3. multiple framing, an appreciation for varied perspectives, and an understanding that more than one thing can be true at the same time, depending on the frame of reference and standards of evidence used; and
4. reflective exploration of meaning, a capacity to think deeply about causal questions and apply this approach when making day-to-day decisions.

The geosciences are well suited to the teaching of these modes of

---

may require consensus across disciplines because they involve complex competing factors that are difficult to reconcile.

thought, Szymanski observed, and service-learning is a way to teach students how these skills apply in public decision making. He involves his own students in projects that engage them in both science and policy.

Szymanski teaches a course called Science in Environmental Policy, which covers the science behind policies such as the Clean Air Act, and also allows students to observe the policy-making process in detail. The course enrolls students from a large variety of majors who bring backgrounds in many aspects of science and business. They learn about the roles science can play in the policy process and consider the competing interests that influence policy outcomes. There is an optional service-learning component to the course, and students who participate in that component earn four credits for the course instead of the standard three. The service-learning students use their business skills to assist a partner organization, such as the American Geosciences Institute, with a specific project related to geosciences. For example, they might research issues, analyze data, evaluate Website functionality, and prepare various types of presentations for their nonprofit partners and for elected officials. Students have taken on such issues as the effects of tax incentives for promoting the use of energy-efficient technologies or products and the difficulties of bringing relevant geoscience information, such as water quality data, to the attention of policy experts.

Participating students travel to Washington, D.C., to present their ideas to their project partners and also meet with policy makers to propose ideas and request action. Szymanski's students have presented research findings in support of policy recommendations, and in turn have seen their findings cited by officials such as U.S. senators. Figure 2-1 shows a poster students prepared to summarize their report on the policy applications of geoscience research. These students made the case for funding basic geoscience research by detailing the relevance of information about energy, climate, and environmental hazards to economic issues and policy decisions. Through these kinds of experiences, the students develop research and other skills, identify barriers, and pursue options for overcoming them. The students also help to boost the influence of knowledge from the geosciences on public policy.

Workshop participants pursued several issues through follow-up questions. In response to one question, Szymanski explained that Bentley has a long tradition of service-learning, but that setting up the geoscience programs nevertheless required some effort, particularly in making the case for the relevance of the geosciences to policy and the importance of transdisciplinary learning experiences. A participant noted that it is important that colleges and universities support and reward junior faculty and graduate students who are willing to sponsor service-learning. Szymanski also noted that he has cultivated professional contacts—particularly non-

**FIGURE 2-1** Policy report poster produced by students.
SOURCE: Szymanski (2016).

profit organizations in need of free analytical work—to help his students find new opportunities. Several comments reinforced Szymanski's point that though service-learning can be difficult, it helps students to cultivate skills and abilities that are easily transferable to other contexts.

## LEARNING FROM EXPERIENCE

Service-learning is gaining popularity within the geosciences but it has a longer history in other disciplines, observed Linda Silka in her overview of lessons from research and experience. She also reviewed research, such as a meta-analysis of studies of the impact of service-learning on academic performance (Celio, Durlak, and Dymnicki, 2011), that has demonstrated the benefits of service-learning for students, institutions, and community partners. The relevant research, she noted, does not always use the term service-learning; she found valuable information by searching other terms, such as "community-university partnerships," "student engagement," "cooperative extension," "science shops," "democratization of science," and "citizen science." Projects that link to the curriculum, allow students to have input into the selection of the issues addressed and regularly ask students to reflect on what they are learning orally or in writing are the most likely to have benefits, Silka found from her review of the research.

Silka observed that, in general, service-learning is a way to increase the depth of students' understanding of the content they study, their capacity to retain it, and their capacity to perceive its application in new contexts. These experiences tend to improve students' attitudes regarding their own learning and their sense of efficacy and also to improve their attitudes regarding the institutions they attend, she noted. These experiences can improve students' social skills and other skills that are of interest to employers and also develop students' commitment to civic engagement. In her paper, Silka elaborated by providing examples of service-learning projects—such as one on the viability of community solar projects and another on the relationship between beehive locations, landscape features, and pollination—that can deepen students' knowledge (Silka, 2016a).

Service-learning's benefits to communities have also been documented, Silka added (e.g., Billig, 2009; Conway, Amel, and Gerwien, 2009). Community partners benefit when students bring fresh, useful knowledge to bear on a problem. Moreover, students may perform tasks, such as data collection and analysis, that the community partners need but for which they lack the necessary staff or resources.

There are several ways to increase the chance that a service-learning experience will help community partners to achieve their goals, Silka explained. Perhaps most important, in her view, is to make sure the project is a true partnership between students and members of the community, one in which each participant has a voice and a role. Such partnerships can improve reciprocity in campus-community relationships. Service-learning should be viewed not as a research project or placement opportunity for students, she commented, but as a collaboration with members of a community. It is also important, she noted, that the students not approach the project with the idea that "we are here to solve your problems our way." Research suggests that if community members perceive that students view their service-learning project as a charitable endeavor or have an arrogant attitude, Silka noted, the result may be more harmful than beneficial to communities and relationships.

Though these features have been widely supported in the available research, Silka noted, there are no "abstract, overarching rules" for how to develop a new program. Examples from other disciplines can provide excellent starting points, she added, but each service-learning program must be tailored to fit the course of which it is a part and the context in which it will be carried out. Identifying community partners and fruitful topics, designing the program, and engaging students are all challenges that require innovation that fits the course, discipline, circumstances, students, and community, she added.

Silka also identified a number of challenges that may concern faculty members who would like to provide service-learning opportunities for their students and offered ideas for addressing several of them; see Box 2-1. Educators with whom she has worked frequently comment that they lack adequate training for their role in overseeing the projects, that they have trouble locating community partners, or that their students do not have the time to participate, for example. She has found that some are concerned about how to structure service-learning programs in the context of a large or online class. Many educators are also concerned about finding topics that will engage their students while also developing deep disciplinary knowledge that can be linked to important issues.

The limited time available did not allow Silka to elaborate on all of the responses to these challenges that she described in her paper (2016a), but she focused on ways to interest students in service-learning. A characteristic of excellent service-learning experiences, she noted, is that their design takes into consideration the needs of community partners, students' interests, and the content of the course. She added that a Venn diagram that identifies the areas of overlap for these three elements is a good analogy for identifying projects that are effective and valuable and will engage students.

"Topics are all around you," she noted, and she used the example of solid waste to illustrate the possibilities. Solid waste problems are expensive for communities, she pointed out. Many are struggling with increasing volumes of waste and waste that is increasingly toxic. The value of many recycled materials has declined, which has complicated efforts in many communities to promote recycling. Silka cited the experience of students and faculty from many disciplines (including anthropology,

**BOX 2-1**
**Common Service-Learning Challenges Reported by Faculty**

Lack of training in service-learning and the challenges of getting started
Identifying appropriate topics for service-learning
Finding community partners
Getting students interested
Tapping into ongoing streams of activities
Identifying what novice students have to offer
Ensuring that service-learning topics help students develop deep disciplinary knowledge
Students with limited time for service-learning
Dealing with large class sizes
Dealing with online classes
Fearing that problems are bigger than what the discipline can offer
Finding, or creating, institutional support for service-learning
Ensuring that habits of service-learning remain with students in their postgraduate lives
Bringing all the project components together

NOTE: See Silka (2016a) for a discussion of these challenges and ways to address them.
SOURCE: Adapted from Silka (2016b).

economics, ecology, engineering, and communications) at the University of Maine to illustrate the possibilities in this topic.

There, students and faculty worked with community partners in the town of Orono to raise awareness of solid waste issues. Students played many roles: gathering information about reusing and composting, analyzing press coverage of waste issues, and working with younger students in recycling efforts. They hosted meetings under the rubric "Talking Trash: The Sustainability Challenge Hiding in Plain Sight" to foster discussion. Local communities with large immigrant populations were especially interested in building public awareness, Silka noted, and students responded with discussion and learning sessions focused on local resources for recycling items that are difficult to categorize.

Silka emphasized that the examples she found might serve as sources of inspiration for geoscience educators—not as blueprints. Her paper discusses ideas she saw in projects from other disciplines (e.g., art, evolutionary biology, ecology, political science), such as ways of approaching potential community partners or bringing together faculty and students from multiple departments to identify opportunities, which could be applied in a geoscience context, For example, in one case she cited, faculty from departments of ecology, engineering, economics, communications, and geographic information systems collaborated to develop a project on the viability and present and future impact of dams in New England rivers (Silka, 2016a).

Silka closed her presentation with some thoughts about resources educators may find in unexpected places. She noted that many campus organizations or institutions have experience with efforts similar to service-learning and also with enterprises that require collaboration among university groups and community members. A national resource is Campus Compact (see Chapter 3). Some institutions also have an office of community outreach or community engagement that could be helpful. Such groups may be in a position to provide advice and examples even if they have not done service-learning projects in the geosciences.

## DISCUSSION: SUCCESSES AND CHALLENGES

Questions and discussion after the two presentations centered on what can make programs more effective. "You need to be willing to make mistakes," said one participant. Another pointed out that it is important that the faculty member involved not "stand above" the community, but that at the same time, the authority that comes with being on the faculty of the local university can be useful. Several comments addressed the importance of collaboration between the students and faculty and the community group or groups with whom they are working. "Real, live partnerships are key," one said, adding that "it shouldn't be the college telling people what they need." A good starting point, another suggested, is to ask a community partner to describe a back-burner project the group would like to do but cannot. Having identified a project the community partner truly needs, students and partners can collaborate to develop the approach that will be most helpful. Another participant talked about the importance of including faculty from other disciplines that may be relevant to the proposed project.

Participants highlighted benefits to students, including the development of social skills needed to work effectively with people who work in different roles, the opportunity to transfer what they have learned to a completely different setting, and the opportunity to pursue an interest that may not have seemed to have an academic application. One suggested using student reflections to demonstrate the depth of the learning that comes with such programs.[3]

The participants were asked to separate into groups to share their

[3]For an example of how reflection is used in science education, see Kober (2015, pp. 6-7).

perspectives on what is being done well in geoscience service-learning programs and areas that still need attention.

## Current Areas of Strength in the Use of Service-Learning in the Geosciences

Participants suggested two areas of particular strength in service-learning programs:

*Service-learning promotes different kinds of learning.* Several participants noted that service-learning projects help students to appreciate the value of nonacademic learning. They may work with practitioners and other people with types of expertise that are completely new to them and deepen their understanding of what they have studied in the classroom. They develop skills likely to be useful throughout life and careers, such as making public presentations and conducting interviews, and they develop attributes such as resourcefulness and resilience. These experiences also introduce students to opportunities they could pursue after college. Participants observed that service-learning experiences can be excellent opportunities for students to experience failure, when their projects are not completely successful and develop the capacity to learn from and build on it. Service-learning projects also have the capacity to engage non-STEM students in geoscience disciplines and to solidify the interest of students who opt to continue in these fields beyond college.

*Service-learning programs engage faculty and communities.* Service-learning can be a learning opportunity for faculty as well as students. Participants noted that many programs have demonstrated the value of listening to community partners and providing services that they need—an experience that can enrich all involved. Some universities have used strategies to explicitly encourage faculty to undertake such projects, including incentives such as fellowships, grants, and official recognition in the tenure review process.

## Areas Where Improvement Is Needed

Participants identified two challenges:

*Students could use more support to help them benefit from service-learning programs.* Several participants suggested that many students could use support in identifying the skills they have that may be useful when working with community partners, and in communicating about those skills to community partners and potential employers. Several suggested that incorporating more opportunities for conscious reflection into projects would help students assess what they have done and learned. Participants also observed that students could use guidance in managing their expectations about the roles they can expect to play, the amount of work that will be needed, and the range of possible outcomes. Faculty could do more to foster relationships that meet students' learning needs as well as the needs of community partners. A related issue is that there are not strong assessment tools for documenting the impacts of these programs for students, a participant commented.

*Institutions could provide more support for faculty.* Many participants offered ways that colleges and universities can support faculty who are interested in service-learning, including providing incentives such as institutional recognition, mentoring by more experienced faculty, strategies to give faculty time for these efforts, and assistance in establishing connections across departments and with community partners. The culture at many institutions is somewhat "risk-averse," noted one participant. There are many barriers that can discourage the pursuit of service-learning programs at many institutions, another commented. It takes years to develop the community ties and structures that are the foundation for the strongest, most stable programs, one noted. Funding structures are barriers to setting up new service-learning programs, another commented, and busy faculty and students can have difficulty making the time for a service-learning project, especially if it requires significant time away from campus. Integrating these programs into courses with high enrollment or in which many students participate online poses logistical challenges, another noted.

Institutions could also help faculty with tracking the benefits of service-learning, both for students and for community partners. Presenting data is very persuasive, a participant noted. Service-learning offers a valuable opportunity to redefine the field, a participant commented, by expanding understanding of the value these disciplines have to offer. "The wall between science and society needs to be taken down," one person commented, and "that involves a big culture change," in which service-learning can play an important part.

# 3

# Service-Learning Benefits for Students and Institutions

Geoscience service-learning can have many benefits for both students and institutions. Two objectives for including service-learning in undergraduate courses are that it will (1) help to engage a more diverse group of students in geoscience fields and (2) help all students who participate to gain skills and other benefits that will last beyond their college years, whether they pursue geoscience careers or not. Presentations and discussions explored available research and other insights on both of these topics.

## REACHING HISTORICALLY UNDERREPRESENTED POPULATIONS

The committee sought two kinds of insight on the goal of using service-learning to increase the diversity of undergraduate geoscience students. Suzanne O'Connell of Wesleyan University discussed the available research on diversity in geoscience fields. Antony Berthelote of Salish Kootenai College and Ben Cuker of Hampton University offered their perspectives based on their experiences teaching at institutions that have long service-learning traditions.

### Research on Diversity in the Geosciences

Questions about diversity in geoscience service-learning are "a wide-open opportunity for research," noted O'Connell, because very few relevant studies are available. There are some data on the numbers of students in historically underrepresented groups who major in geoscience-related fields, O'Connell noted, which show that the geosciences are "no longer the worst of the worst" in terms of the diversity of their students. Figure 3-1 shows the numbers of students who are members of population subgroups (American Indian/Alaskan native, black, Asian/Pacific Islander, and Hispanic) who earned geoscience bachelor's degrees from 2002 to 2012. Total numbers across the four groups increased during this period, with the largest gain among Hispanic students. Students in these groups were 2 percent of all bachelor's degree earners in 2002 and almost 18 percent in 2012, O'Connell added. The numbers of geoscience Ph.D.s awarded across the United States to individuals who belong to racial or ethnic minority groups more than doubled in the same time period, O'Connell added, though overall numbers are low. There were fewer than 20 Ph.D.s awarded in 2002 and 55 in 2012 (National Science Foundation, 2012).

A few researchers have also shed some light on why participation has been low for some groups, O'Connell noted (Whitney et al., 2005; Huntoon and Lane, 2007; Levine et al., 2007; Hoisch and Bowie, 2010;

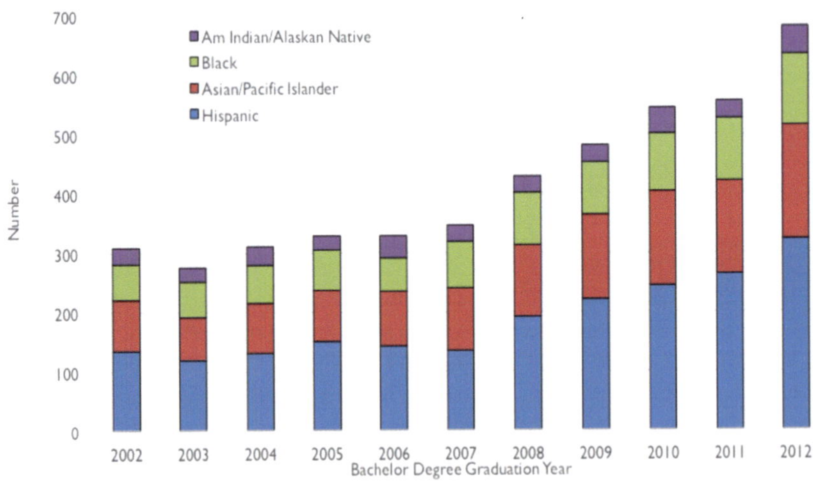


**FIGURE 3-1** Recipients of bachelors' degrees in geoscience fields by underrepresented groups, 2002-2012.
SOURCE: National Science Foundation and National Center for Science and Engineering Statistics (2015).

Velasco and Velasco, 2010; Stokes, Levine, and Flesse, 2015). This work suggests, she explained, that four factors tend to deter some students from enrolling in the geosciences: lack of exposure to the geosciences, lack of interest in or exposure to outdoor activities, interest in a career with greater perceived prestige and perception that the geosciences are not relevant to their interests. These factors apply to all students, she added, and they suggest insights for increasing diversity.

One problem is that many students have not been exposed to geoscience fields and are not fully aware of the range of employment opportunities for graduates. To address this problem, O'Connell suggested, geoscience faculty and other geoscientists interested in attracting more students to their fields might look for opportunities to make their work better known in their communities. Geoscience departments can also do more to publicize available data about hiring and graduation rates for their students and starting salaries. Department Websites, for example, could easily post such information, she noted.

Opportunities to do field work have long been a draw for students who have grown up with considerable exposure to the outdoors, O'Connell noted. These experiences have been comparatively more common for white young people than for those in underrepresented groups, however. To broaden the appeal of the geosciences, she suggested, it will be important to help prospective students see that field work is not an essential component of study or employment in these disciplines. Geoscience departments can do more to show that careers may also involve laboratory and computer work, and working with people. An additional problem is that some students perceive geoscience careers as having limited prestige. Departments can address these perceptions, O'Connell pointed out, by using their Web pages and other marketing strategies to highlight the many contexts for geoscience work, the range of careers their graduates have pursued, and the contributions of well-known geoscientists. O'Connell said she sees geoscience majors as an excellent foundation for other careers, including medicine, law, and politics.

A fourth issue is that prospective students may not readily see the ways in which the geosciences are relevant to their own lives, O'Connell explained. Service-learning projects can help them experience directly how these disciplines contribute to such vital issues as climate change, sea-level rise, challenges to the water and energy supply, and land use.

O'Connell observed that qualitative research has shown that service-learning benefits students by increasing their sense of personal efficacy, awareness of the world, their own personal values, and engagement in their classroom experiences (e.g., Astin et al., 2000). However, the research overall leaves many questions about the students who do and do not enroll in geoscience courses and the role service-learning can play in attracting a more diverse student population.

O'Connell and her colleagues at Wesleyan University have found that opportunities to reflect on their learning help both students and faculty to deepen their experiences, and this is particularly true for service-learning projects. They have used an approach called Describe, Examine, and Articulate Learning, O'Connell explained, to help students reflect on their learning (Ash and Clayton, 2009). O'Connell believes that recruiting is key at every level: She is constantly looking for students who might be interested in postgraduate study, for example. Service-learning projects can be an excellent way to encourage students to continue their geosciences studies, she commented in closing.

## Views from Two Institutions

### *Salish Kootenai College*

Berthelote reflected on his experience at Salish Kootenai College, which serves primarily Native American students. He agreed with O'Connell that lack of exposure to the geosciences is a key reason why relatively few students from traditionally underserved population subgroups enroll in courses in these disciplines. He suggested that cultural differences may partly account for students' varying attitudes with respect to geoscience fields. Another factor he has observed at his own institution is that many students are in need of remedial classes in mathematics and science and may view requirements for enrolling in geoscience classes as an obstacle.

The geosciences have "a PR [public relations] problem," Berthelote suggested, in that their potential alignment with cultural values that have a high priority in Native American and other minority traditions is not always well understood. In his view, Native American students may place a particularly high value on community and respect for the planet, in comparison with members of other groups.[1] Service-learning is a critical way to help students from underrepresented groups who bring varied cultural experiences to their undergraduate studies to see study of the geosciences as an opportunity for community empowerment and for contributing to the protection of the Earth, he suggested.

Berthelote described one approach to promoting community service, civic engagement, and service-learning in higher education, the Campus Compact. This is an association, initiated in the 1990s, through which approximately 1,100 colleges and universities in 34 states have made a

[1]Berthelote referred to Gregory Cajetes' 1999 book, *Igniting the Sparkle: An Indigenous Science Education Model*, as background for this discussion.

commitment to developing their students' civic engagement and dedication to service. The compact offers resources to support educators and staff in providing opportunities for students.[2]

Salish Kootenai College, Berthelote noted, is a member of the Montana Campus Compact and has incorporated the compact's view of citizenship into the four primary learning objectives for its students (the other three are critical thinking, communication, and culture). The college defines citizenship as "informed and committed participation in the life of one's community at the local, national, and global level," he explained. As these four objectives were integrated into academic programs, he added, service-learning was included as a degree requirement for every major. Students are required to participate in at least 30 hours of service that takes place outside of regular course time, he explained, pursuing objectives tied to course content.

Service was recently made an explicit part of the college's formal vision statement, Berthelote added, because it has become a key element of the student experience:

> Salish Kootenai College aspires to be the pre-eminent educational center of excellence for American Indian students, grounded in the cultures of the Sélis, Ksanka, and Ql̓ispé people of the Flathead Nation. The college will empower students to improve the lives of their families and communities through research, leadership, and service.

"Undergraduate minorities will flock to the earth sciences" when they see them as a means to serve their communities, Berthelote concluded.

### *Hampton University*

Hampton University, Ben Cuker explained, was founded in 1868 with the mission of serving newly freed slaves in the Hampton Roads region of Virginia. The site where the university was founded had been a destination for approximately 50,000 slaves who took refuge there and obtained their freedom. The federal agency formed to aid former slaves, the Freedmen's Bureau, set up many schools and colleges, of which Hampton University was one. The school's mission—to prepare its students to take a full part in American society—focused on development of "the head, the heart, and the hands," Cuker explained. "That is service-learning," he added. In the early years the students not only learned academic subjects, but also raised their own food and built and maintained the college build-

[2]See http://compact.org/ [May 2016]. Berthelote also pointed participants to the University of Southern California site mentioned earlier, at https://dornsife.usc.edu/history-of-service-learning/ [August 2016], for background on service-learning.

ings. Thus, service-learning and sustainability are integral aspects of the school's heritage, though they did not always go by those names.

As Hampton University was growing, Cuker went on, changes in the philosophy of education were influencing the structure of academic departments and objectives for learning. An "industrial model," in which the emphasis was on preparing students for work, began to supplant a broader intellectual approach that had guided many institutions, Cuker suggested. Disciplines were increasingly taught in "silos" as this model took hold, he continued, and one consequence was a growing demarcation between applied and "real" science. The connection between knowledge and practical applications was de-emphasized. Today, however, he suggested, college students are very interested in connecting what they learn with real-world issues.

Students tend to come to Hampton University with a strong focus on their role in the community and are very amenable to service-learning projects. These projects benefit all types of students, in Cuker's view. He and his colleagues have assessed the results of these projects and their impacts on students, but have attempted to do so with a minimum of paperwork. It is important to embrace "other ways of knowing," he added, which cannot always be measured on a metric scale.

## Discussion

Committee member Sue Ebanks highlighted key points from the discussions of diversity. "Perception is a big deal," she noted, adding that it appears that there is a widespread misconception that the geoscience community is a closed society that is not welcoming to underrepresented students. Improved marketing is a promising strategy for reaching these students and helping them understand the benefits and opportunities in the geosciences, but new approaches are needed to invite in more students from varied backgrounds, she noted. A diverse faculty, like that at her own institution, Savannah State University, can be extremely helpful in figuring out effective strategies for reaching these students, she added.

Geoscience programs could significantly improve their marketing, Ebanks added. Students are likely to have a range of both positive and negative impressions of the sorts of work that geoscientists are involved with, such as the clean-up efforts associated with the Environmental Protection Agency, preservation of the environment, or the opportunity to do field work outdoors, Ebanks noted. Highlighting the full range of possibilities that are available in these fields—including indoor opportunities such as bench science and the chance to address urban issues—can also help draw students in, she noted. "To be relevant," Ebanks continued, "you have to know the audience" and address the concerns of the students the program would like to attract. This challenge goes beyond marketing, she continued, noting that "we need to ask students what they are interested in," she said, and "listen to their answers."

Participants followed up on several of these points. Several pursued the question of marketing the geosciences. "We've known we have a PR problem for more than 20 years," one commented. The fact that there is still a problem may be signaling that students need more than marketing strategies to show them how the geosciences can be relevant to their lives and incorporate activities they truly want to engage in. Service-learning is an ideal opportunity for students to experience the possibilities for themselves and see if they are interested, this person suggested.

Another participant commented that progress in increasing the diversity of geoscience students has lagged "because we haven't changed how we do things." Waiting until students get to college is too late, another person noted. Recruiting students for an introductory college class is critical, but it is challenging when they have many competing interests and limited experience with the range of what the geosciences offer. Elementary school is not too early to begin, another person noted, and service-learning projects that involve younger students are excellent ways to "work on the pipeline."

Another set of comments pointed to the key role service-learning can play in showing students the values that are inherent in the geosciences. "The message regarding sustainability is really not out there" one person commented, adding that service-learning can address students' interest in directly helping their own communities.[3]

## PREPARING STUDENTS FOR EMPLOYMENT

Service-learning can have many benefits for students that last long after the program is completed. John Gierke of Michigan Technological University and the Peace Corps Master's International Program,[4] Amy Cohen of George Washington University, and Eric Riggs of Texas A&M University offered their thoughts about the connections between service-learning and employment. They drew on their experiences in research and in service with such organizations as Learn and Serve America at the Corporation for National and Community Service, the American Geosciences Institute, and the National Association of Geoscience Teachers.[5]

[3]This participant referred participants to David Orr's 2004 book, *Earth in Mind*, for a detailed discussion of this point.

[4]Shortly after this workshop, the Peace Corps announced the end of its Master's International graduate school programs.

[5]Materials shared with participants before the workshop and available at http://sites.nationalacademies.org/DBASSE/BOSE/CurrentProjects/DBASSE_169060 [October 2016]

Participants then separated into groups to discuss ways service-learning can help students be more competitive in the workplace and summarized their ideas.[6]

## Evidence of Impacts on Employment

Gierke, Cohen, and Riggs were asked to address four questions.

1. *Is there evidence that service-learning helps students get jobs?*

There is considerable anecdotal evidence that service-learning experiences help students gain employment, in Gierke's view. Employers tend to value skills that have been developed in a real-world context, he added, and there is some indication that among employees at federal agencies, people with these experiences tend to progress faster than others.

Cohen concurred and added that there is some research showing that students who have been involved with service-learning or other volunteer activities tend to find jobs more quickly than students who have not participated in service-learning and also to earn slightly more in their first jobs (Eyler et al., 2001). She suggested two possible explanations. First, these experiences can help students build social capital—professional contacts, job leads, and social relationships—that help them secure employment and progress once hired. Second, they also develop human capital; that is, students gain knowledge and skills, have leadership opportunities, and work experience that all help them to be successful in interviews and in the workplace. Cohen has often observed students finding jobs with organizations that they had worked with as undergraduates, she added.

Riggs' answer to the question was a "qualified maybe." It is very difficult to distinguish the effects of earning a geoscience degree from the effects of service-learning experiences students have while pursuing such a degree, he pointed out. He agreed that developing networks can be very helpful in finding employment, and that service-learning can shape students' aspirations and career development. In his view, however, it is difficult to establish a causal relationship between service-learning and later outcomes.

2. *Does service-learning help create a sense of professionalism among students?*

Gierke said he was "floored" by the professionalism of the students

---

were chosen to provide background for this session. See, in particular, Battistoni and Longo (2006).

[6]Background materials for this topic are posted on the project Website at http://sites.nationalacademies.org/DBASSE/BOSE/DBASSE_171829 [September 2016].

who participated in the Washington trips Szymanski had described (see Chapter 1). He said he had rarely seen students performing at quite that level, but that he has seen students demonstrating improvements in efficiency and planning and the capacity to handle travel and other logistics. They develop confidence through the "school of hard knocks" that can be an aspect of service-learning, he added. He also said service-learning activities "are contagious" in that they tend to attract able and energetic students to geoscience departments.

Cohen agreed that service-learning can have this effect, but she cautioned that the benefits come when the program is designed specifically to develop them. "If you have the intention to develop writing and communications skills," she suggested, "you are more likely to achieve that result." Among the most common outcomes she has seen are improvements in the ability to cope with strangers, team work, communication and presentation skills, and the capacity to draw on expert knowledge to solve problems. It is important that students be encouraged to reflect on what they have learned, she added, to help them solidify those benefits.

Riggs has also observed students developing skills that can help them in professional settings, though he noted that collecting evidence to document such benefits is tricky. Nevertheless, he has seen students improve in "cultural competency": They learn to behave appropriately in different environments, how to "read people" and not to assume that the scientific knowledge they bring is necessarily superior to other kinds of knowledge that are relevant to the project. Opportunities to write about science topics in nonscience settings, or to address social and political issues in a science context are all ways that students broaden their understanding, he added. As they check the knowledge and understanding they gained in the classroom against real-world challenges, he added, they develop in ways that cannot easily be taught in the classroom.

3. *Is there any evidence that service-learning can help create a more positive impression of a college or university within the local community?*

Gierke said that he had seen some indications of possible benefits to an institution's reputation but did not suggest that any evidence is available on this question. Cohen agreed that it is difficult to find empirical evidence of benefits. She noted that surveys of community partners may not be very informative, since their staffs are usually very grateful for all assistance they receive and tend to rate the experiences positively regardless of the outcomes. It can be challenging to find objective sources of data to use in the evaluation of service-learning projects, but indications of a project's impact might include informal feedback or actions taken by a local government entity following a service-learning project. In general, she has observed that service-learning is perceived as helpful.

Riggs agreed that general perceptions are positive, but that the nature of the feedback depends in part on which communities are asked. Some groups within communities may feel ignored, he noted, adding that service-learning projects that engage community members in a collaborative way may counter negative impressions of the college or university. If students and faculty invest time, they will get trust and access from the community partner in return, he added. Relationships between schools and communities need to be maintained over time, so perhaps the best evidence of success is reciprocal engagement that is sustained.

4. *What can be done to create more service-learning opportunities?*

Capstone experiences that require service-learning for geoscience majors are "a natural fit," commented Gierke. He advocated making such experiences a requirement for accreditation—like the requirement that engineering programs have capstone experiences for students—and that institutions provide resources to support it. The resources needed include faculty time as well as funding to cover expenses such as travel costs, so institutions and departments would need to consider ways to make this work. It would also be helpful, he added, if institutions could streamline the administrative process so that obtaining approvals and other necessary steps were not obstacles for students or faculty. Gierke would also like to see institutional leaders articulate the value of service-learning to help inspire faculty to promote it.

Cohen added that involving more potential employers would be a good way to boost participation. Nonprofit employers would likely fit better with academic institutions, she acknowledged, and any such relationships would help to "get the word out" about the benefits of service-learning. It can take long-term effort to form productive relationships with businesses, agencies, and institutions, she noted.

Riggs noted that many faculty are already organizing activities that are close to meeting the criteria for service-learning, and that it would be relatively easy to adapt them so they do meet service-learning objectives. There are few human enterprises that do not involve the geosciences in some way, he noted, so there are many cross-disciplinary opportunities. He also noted that involving graduate students and preservice teachers in undergraduate service-learning projects increases the possibilities for spreading the word: Participants who tell others about their own activities may inspire potential students. Weaving service-learning into the undergraduate experience would be easier, he added, if institutions modified institutional procedures to facilitate these experiences and awarded credit for service-learning.

## DISCUSSION: IDEAS FOR ENHANCING STUDENTS' COMPETITIVENESS IN THE WORKFORCE

Participant discussions covered three themes: the outcomes of service-learning that can benefit students and institutions, ideas for measuring those outcomes, and implications of these points for the design of service-learning programs.

### Possible Outcomes for Students and Institutions

Participants identified learning outcomes for students that help them develop in their careers and thus benefit their institutions as well.

*There are intellectual benefits for students.* Students may learn to apply the tools, habits, and knowledge of the geosciences to societal problems. The experience of dealing with uncertainty and complexity or with ethical issues, as well as the chance to see how material they have studied can be applied in the real world, are all likely and important outcomes.

*Students develop increased professionalism.* Service-learning provides students with the chance to lead and manage projects. They learn from seeing models for designing and running a project that can be applied in other situations. They also develop self-confidence and self-efficacy and may begin to view themselves as professionals. The dedication and effort required for service-learning projects are recognized by many employers, and students also gain access to resources and networks through their involvement with organizations outside the college or university.

*There are numerous benefits for institutions.* Service-learning programs may have positive effects on the relationship between the school and its community. Participants suggested that the effects are most likely to be positive if the faculty members involved understand the history of the relationship, paying attention to negative incidents that may have damaged it. Faculty should also communicate clearly with community members (beyond the partners directly involved) about the nature of the project they propose, who will be involved, where it will take place, how it will work, and what its goals are. They should be specific about how possible concerns, such as the privacy and safety of individuals involved, including the students, for example, will be addressed. Expectations—including the possibility that the project will not be completely successful—should also be addressed from the beginning.

### Assessing Outcomes

Using data to evaluate and improve programs and courses is important, several participants noted. Data can be used for spreading the word about their benefits and generating support within institutions and

**TABLE 3-1** Example Assessment Models Suggested by Participants

| Assessment Ideas | Link |
|---|---|
| Many assessments can be aligned with existing communication, team, and project-based learning | http://serc.carleton.edu/introgeo/assessment/index.html |
| Assessments that consider collaboration and relevance to partner | http://serc.carleton.edu/NAGTWorkshops/servicelearning/activities/134864.html<br>http://serc.carleton.edu/NAGTWorkshops/servicelearning/courses/127425.html<br>http://serc.carleton.edu/NAGTWorkshops/servicelearning/activities/127457.html |
| Implementation tips: SERC Service Learning Assessment<br>SENCER tools | http://serc.carleton.edu/introgeo/service/howto.html<br>http://www.sencer.net/Resources/models.cfm |
| Other Resources | http://serc.carleton.edu/departments/index.html<br>http://serc.carleton.edu/NAGTWorkshops/departments/degree_programs/matrix.html |

NOTE: All above links were accessible as of January 2017.

beyond. Using data to demonstrate benefits are invaluable in making the case for allocating resources or expanding a program, and it is most useful to have information about both learning goals for students and project outcomes.

The assessment tools used for most academic courses will provide useful information about many outcomes, participants noted. Some outcomes that are particular to service-learning experiences are difficult to measure, but participants suggested possibilities and identified some models that could be useful examples (see Table 3-1). One possibility, several participants suggested, is that assessments could be used at the beginning and end of a program to measure growth in technical skills or civic knowledge. Surveys and site visits by faculty could be used to solicit feedback from community partners. Surveys and assigned reflection papers could collect information about students' attitudes and perceptions of what they learned from the experience. Professional skills, such as communication, leadership, and management, as well as other objectives such as increasing awareness of diversity, may be more difficult to assess, it was noted.

One breakout group suggested possible outcomes for students (see Box 3-1) and also suggested possible ways of assessing the kinds of objectives that are specific to service-learning:

- Site visits for quality control, such as checking progress and proficiency with skills or tools
- Surveys of community partners
- Progress checkpoints for documentation of incremental progress;
- Project preparation obligations, such as reading assignments
- Peer review of products before they are released to partners, with special attention to alignment of product with community partner needs
- Review of assignments from partner information sources (fact sheets, Websites) or other relevant sources, such as local news or scholarly literature
- Assignments that explore data collected from geoscience databases (e.g., local watershed, climate, or other data) that students will use to understand trends important to framing the problem
- Reflection activity in which students characterize topics, such as their own role in solving problems, or their role in lifelong stewardship and decision making, where evaluation criteria could include incorporation of examples or writing quality
- Reflection in which students consider issues that might be sensitive in the community before a site visit (to guide them to present what they are undertaking in ways that are not offensive to the audiences they serve)
- Assessment activities embedded—designed to fit into—the stages of the project. For example, students might be asked to write a piece about the diverse audiences involved in the problem, or to explicitly incorporate diverse disciplinary perspectives into their documented introductory research or in a class discussion
- Assessment of leadership based on the level of engagement of the audience. For example, if students design a project to reach or support particular groups, individuals who engage in the project could be counted or surveyed. Community partners or mentors could also be asked to evaluate students' performance at, for example, sharing initial project ideas and soliciting early feedback, or describing and supporting new ideas or changes
- Literacy or competency assessments aligned with particular disciplines (e.g., national literacy standards). Faculty could modify these to suit the service-learning context, by adding challenges

such as characterizing the importance of the issue addressed by the service-learning project and its relevance at the local, national, or global scale, or discussing trends in relevant data
- Student demonstration of understanding of systems by diagramming the kinds of interactions that played a role in the project (e.g., interactions between human processes and environmental cycles)
- Evaluation of online forums through which project partners interact

## Implications for Program Design

The design of service-learning projects should take into account both goals for student learning and goals for meeting community partners' needs, several participants observed. They identified several elements of

### BOX 3-1
### Service-Learning Objectives that Could Be Assessed, Suggested by Participants

**Communication/Science Literacy**

1. You will be able to communicate geoscience ideas orally and in writing, in ways that are effective for the audience you serve (e.g., partner, public, campus).
2. You will evaluate sources of information for reliability.
3. You will identify central community concerns that tie into your geoscience topic by reviewing local news sources or social media.

**Applying Knowledge**

4. You will be able to apply the tools, habits, and skills of a geoscientist to a societal problem.
    a. You will synthesize information to frame a problem of relevance to the community partner served.
    b. You will gain confidence in describing trends or rates of change using graphic representations to build understanding of the service-learning project.
    c. You will apply quantitative or spatial analyses or modeling tools to data relevant to your partners.
    d. You will learn the field methods critical to evaluating a societal problem.
    e. You will learn the analytical strategies used to evaluate a geoscience problem in context of society.

**BOX 3-1 Continued**

**Collaboration**

5. You will contribute to the understanding of a societal problem and report and reflect on your contribution to team problem solving.
6. You will meet all responsibilities on your project timeline and maintain a reliable connection with your partner and team.

**Becoming Proactive**

7. You will develop solutions for your project partner(s) that are aligned with their needs.
8. You will design a product (e.g., policy recommendation, fact sheet, or Website) that describes proactive strategies for your partner or the audience they serve.
9. You will describe how your participation influences your future decision making.

**Diversity/Justice**

10. You will incorporate the views and ideas of the audience you are serving into your service-learning project.
11. You will incorporate diverse perspectives of contributors into your service-learning project.
12. You will be able to describe the diversity of disciplines, ideas, and people needed to address the problem.

**Sustainability**

13. You will be able to describe how people's use and management of earth resources are directly tied to responses in spheres of sustainability (environmental, societal, economic).
14. You will describe (or diagram) connections, fluxes, fates, and feedbacks in the system you are evaluating.
15. You will describe how your service-learning project relates to a larger challenge.

**Reflection**

16. You will reflect on the relevance of your work to the community/partner you serve.
17. You will reflect on the relevance of your work to your future decision making or behaviors.

**Leadership**

18. You will lead an activity that engages a new audience.
19. You will present a product you create for a community partner/audience.

strong design. Formally involving graduate students and others as *mentors* to undergraduate participants was identified as very useful. Mentors are not only a resource but also a conduit for expanding outreach. The experience of serving as a mentor is also valuable for graduate students, participants commented, though they also need support in finding time for these efforts. Attention to the *professional development* that mentors and faculty may need to succeed is also important, and research on the roles they may play could be used to improve this kind of support, it was noted. Participants also advocated scaffolding, or identifying and describing the nature of and goals for the stages and elements of the project, is likely to make it easier for students to understand what is needed and to enhance their effectiveness. *Support from the institution* was also identified as a key ingredient of success, and participants advocated that this be built into the plans to the extent possible. *Plans for assessing outcomes* should also be integral to the project design, several participants emphasized. Information collected should be used to evaluate and improve both courses and programs.

# 4

# Characterizing Current Service-Learning

The discussion turned next to the nature of service-learning experiences that are currently available in geoscience disciplines. An overview of research and other information provided a look at the nature and reach of these activities, and this was followed by a detailed look at two ways of incorporating service-learning into undergraduate geoscience courses. Participants then broke into groups to reflect on the primary elements of service-learning from the perspectives of different geoscience disciplines, types of institutions, and types of students.

## WHAT DOES SERVICE-LEARNING LOOK LIKE?

Suzanne Savanick Hansen of Macalester College and committee member Sarah Fortner of Wittenberg University reviewed the available research and analysis that bears on the implementation of service-learning in the geosciences. They looked for information about the types of activities that are used in such programs, information about desired goals and outcomes, best practices that have been identified, and resources such as handbooks and toolkits.

Savanick Hansen began by cautioning that the review she and Fortner conducted was not a comprehensive analysis but a synthesis of available resources. They reviewed a preliminary literature survey conducted by staff, explored examples from the Science Education Resource Center site and Campus Compact, and reviewed abstracts from a Geological Society of America (GSA) conference to develop a broader sample of the coverage of service-learning. Savanick Hansen identified some themes from this literature.

With respect to the variety of experiences available to students, she observed, the material they reviewed addressed programs that may or may not have been identified as service-learning. These programs did, however, involve key elements of service-learning, such as interdisciplinary research, real-world problem solving, and development of skills likely to be useful to students after they begin their careers. Many gave students the opportunities to develop informational materials using geoscience research and present them to community members, but not all included an opportunity for reflection. Figure 4-1 illustrates the kinds of student experiences that overlap with service-learning.

Savanick Hansen noted that the service-learning projects they found generally fell within four areas of the geosciences: geology (hydrology/

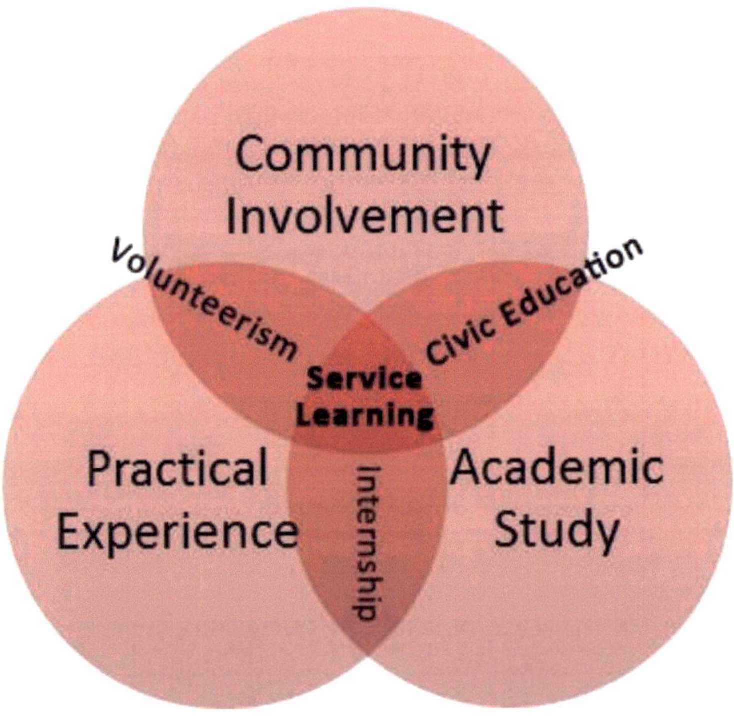


**FIGURE 4-1** Student experiences that overlap with service-learning.
SOURCE: Youth Rebuilding New Orleans (2015). Reprinted with permission.

**FIGURE 4-2** Frequency of terms found in service-learning-related abstract titles, 2001-2015.
SOURCE: Savanick Hansen and Fortner (2016).

water quality, soils); atmospheric sciences (climate change); oceans; and the polar sciences. Issues related to sustainability are a particularly common focus, and these programs are likely to be interdisciplinary and to involve applied research. They created a word cloud (a computer-generated image) to depict the frequency of terms used in these abstracts; see Figure 4-2.

The learning outcome that was best documented in the literature, Savanick Hansen went on, was an increase in students' engagement with the academic material they were studying. Students also improved in technical and other skills, such as communication, collaboration, project management, and the capacity to be aware of and reflect on their own learning. Community partners also generally reported receiving benefits from their involvement in service-learning programs.

Savanick Hansen highlighted a few examples that illustrate points in the literature. One finding was that geoscience service-learning projects are particularly likely to deal with water and soil quality, oceanography, atmospheric science, and the polar region; see Table 4-1. Arsenic on Main Street, for instance, was a water quality fair held in Unity, Maine, at which students analyzed water samples for residents. In another example, a course at Macalester College, students collected data on greenhouse gas

**TABLE 4-1** Geoscience Service-Learning—Sample Programs

| Program | Information |
|---|---|
| Arsenic on Main Street | http://serc.carleton.edu/NAGTWorkshops/servicelearning/activities/39589.html |
| GLISTEN (Great Lakes Innovative Stewardship Through Education Network) | http://ncsce.net/glisten/ |
| REU on Sustainable Land and Water Resources | https://reuslawr.wordpress.com |
| Monitoring Lead in an Urban Community Garden | http://serc.carleton.edu/NAGTWorkshops/servicelearning/activities/135228.html |
| Marine Environmental Geology | Bowdoin College Course |
| Campus Greenhouse Gas Emissions Inventory, Climate Action Plan | Macalester College Course |
| Local Solutions to Global Climate Change | Wittenberg University Course |
| Juneau Icefield Research Program | http://juneauicefield.com |
| Local Hydrogeology in the Classroom | http://serc.carleton.edu/NAGTWorkshops/hydrogeo/activities/10042.html |
| Wastewater Treatment for a Polluted Harbor | http://serc.carleton.edu/NAGTWorkshops/servicelearning/activities/39020.html |
| Investigating Contaminant Transport with Local Sierra Club | http://serc.carleton.edu/NAGTWorkshops/servicelearning/activities/39016.html |
| Ecotoxiocology | http://compact.org/resource-posts/ecotoxicology/ |
| Geoscape of Bozeman, MT (intro level example) | http://serc.carleton.edu/NAGTWorkshops/servicelearning/activities/39375.html |
| Bridging Undergraduate Learning, Teaching and Research (advanced level) | University of Connecticut |
| Oceanography Beach Cleanup | http://serc.carleton.edu/NAGTWorkshops/servicelearning/activities/129095.html |
| Hydrology Service Learning | University of Vermont |
| Fostering Our Community's Understanding of Science (FOCUS) | http://compact.org/resource-posts/focus-fostering-our-communitys-understanding-of-science-service-learning-experience/ |

NOTE: All above links were accessible as of January 2017.

emissions and used these data to help the college develop a climate action plan.

Fortner described another example that demonstrated some of the benefits of service-learning to students and community partners. In an introductory geology class taught at Wittenberg University, she explained, students learned about climate change issues from community experts—such as watershed managers and facilities personnel—and produced learning modules in collaboration with experts at a local science museum. Students reported in surveys that the experience had increased their awareness of the relevance of climate issues in many fields, she noted. Several also benefited from ongoing relationships with community partners after the course by doing internships and other activities. The museum developed the model for use in training graduate students in science communication.

At Wittenberg University, Fortner went on, community-based research has become an explicit focus in geology and environmental science courses. The university has documented benefits of this type of research for students, including research experience; access to resources, internships, and other opportunities; and increased understanding of the real-world relevance of their academic studies. Faculty have benefited from the opportunity to collaborate with colleagues and community partners, and those partners have reported improvements in their environmental decision making and management and increased public awareness of their work.

Public outreach about relevant science issues is a particular hallmark of programs that address climate change in polar regions, Fortner explained, and the Juneau Icefield Research Program—a science and education program initiated in the 1940s in which students and scientists live for 2 months in the Juneau Icefield to conduct glaciological research—is an example of that. The students receive training in public communication and share what they learn both in a public blog and in a presentation to the community.

Fortner noted that service is highly valued by many colleges and universities as a marketing tool. Institutions recognize the benefits for students who participate, but there is little evidence that institutions demonstrate the value they place on these programs by providing tangible support and rewards for the faculty involved. For example, interdisciplinary service-learning projects require extra time—for developing partnerships, locating funding, and getting the resulting work published—all critical for earning tenure. Collaboration that involves policy makers and the public adds to the complexity of the challenge of institutional arrangements, she noted.

Despite the challenges, Fortner explained, she and Savanick Hansen found from their review of GSA abstracts that interdisciplinary efforts have become more common in the geosciences, and that the focus on applied problem solving has increased. Fortner concluded by reminding the group that the primary message from anecdotal evidence, course evaluations, and reflections is that these experiences tend to increase students' engagement with the material they are studying.

## TWO CURRENT APPROACHES

Caroline Davies of the University of Missouri, Kansas City, and Bob Gilbert of the University of Texas, Austin, each described ways they use service-learning in their own teaching.

### Service-Learning at the University of Missouri, St. Louis

Davies includes service-learning in all of her classes, she noted. Her four goals for these experiences are that they:

1. emphasize the integration of geoscience and social challenges,
2. engage students in hands-on interactions with the community involved,
3. develop students' science communication skills, and
4. develop students' skills as "change agents."

She used an example from an upper-level class in environmental sustainability to illustrate how she pursues these goals through a required community engagement project. The course content for this class changes from year to year, Davies noted, but the students have the opportunity to select, design, and implement their own projects. In a recent year, the course focused on urban agriculture. The students worked with community members on plans to grow food in their neighborhoods. Davies engaged colleagues from within and outside the institution with expertise in environmental urban planning, historic preservation and community development, farming, and the geosciences to collaborate with her in encouraging the students to think broadly about the issues associated with urban farming. The students worked with community members to develop neighborhood assessment plans for growing food that included assessment of the environment before and after a garden was installed.

In another project, students developed a plan to help the University of Missouri reduce its energy consumption by installing occupancy sensors that activate lighting only when it is needed. The students found that installing the sensors would cost about $7,000 but save the university $13,000 annually. The poster they used to present their findings is shown in Figure 4-3. Other students tackled topics such as affordable sustainability practices for a local YMCA and a local prison, a textile recycling drive, and a campus-wide forum on improving sustainability within the university. Each year students present their final projects at a juried symposium, in which city officials, architects, neighborhood leaders, and others participate.

The students' reflections illustrate the enthusiasm many of them feel for their projects and their appreciation for what they had learned about sustainability, Davies noted (see Box 4-1). Since 2008, she added, the number of students participating in these projects in a given year has grown from 80 to 410, and the number of projects has grown from 35 to 327.

### Service-Learning in an Engineering Course

Gilbert identified key objectives he has for a required senior-level civil engineering course, one that engages students in geotechnical engineering:

- Apply ethical reasoning and leadership in the practice of engineering.
- Discuss the roles of science, technology, and the needs of society in conceiving engineering solutions.
- Understand the importance of leadership in successful project implementation.
- Explain the need for lifelong learning.
- Give examples of the legal and business aspects of professional engineering practice.

He noted that these are among the most important goals for professional engineers, and yet are rarely taught well at the undergraduate level. Gilbert and his colleagues had found that a straight lecture format was not effective at preparing students in these areas, so they converted the senior course to a service-learning model. Each year the students tackle a new set of projects that meet genuine needs in the community and may be implemented. The faculty turned to the city of Austin, as well as the university's facilities staff, to identify local needs that the students could address. Recent projects have addressed, for example, trails and sidewalks, rain gardens, and accessibility for people with disabilities.

For each project, Gilbert explained, students work in teams of three or four, with the guidance of volunteers from the Austin branch of the American Society of Civil Engineers—in 2016, 40 mentors worked with the students. Each team develops a plan to solve a particular problem and presents it, both orally at a public meeting and in writing to the clients who had requested the assistance. In 2016, 93 students were involved,

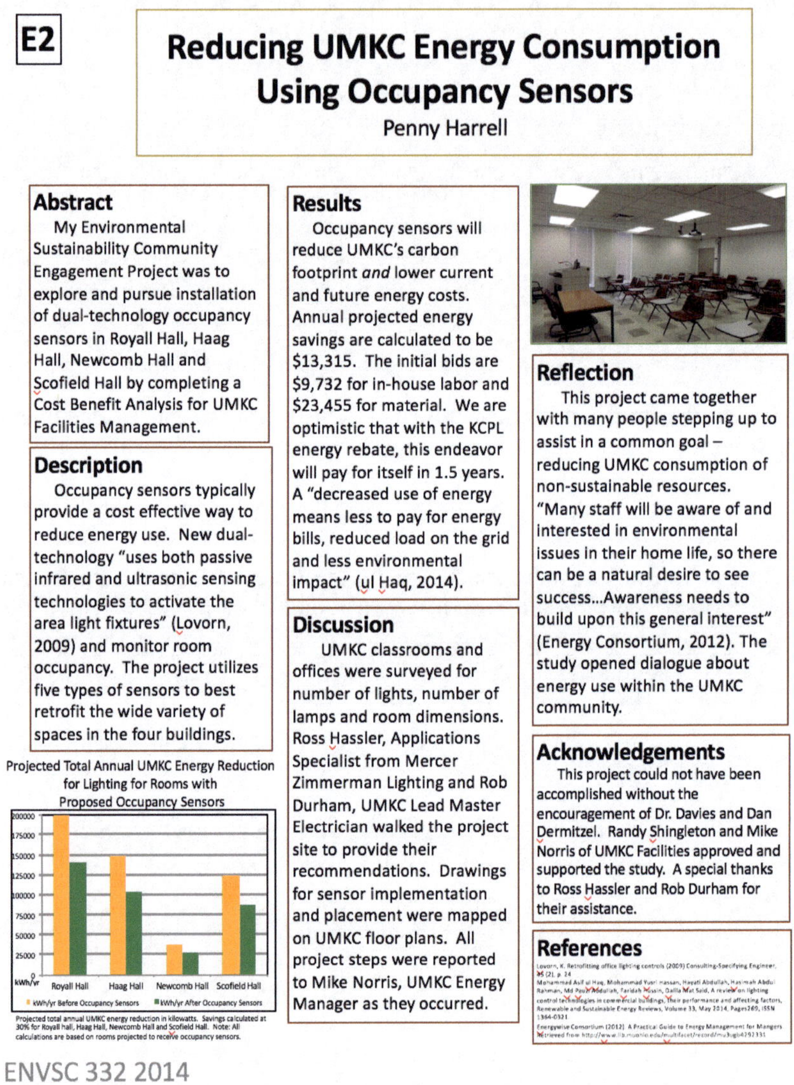


**FIGURE 4-3** University of Missouri, Kansas City, student presentation of an energy reduction plan.
SOURCE: Davies (2016).

**BOX 4-1**
**Sample Student Responses to a University of Missouri, Kansas City, Service-Learning Project**

The class really inspired me to look outside the norm and see what could be instead of what is . . . It has been hands down my favorite project in my college career. . . . The idea that our designs could possibly be implemented in exciting.

I have learned so much about urban agriculture and what it's all about. Another concept is sustainability, I thought I had an okay idea of that before coming into this class, but now I feel like a have a great handle on it.

It was honestly life changing for me and I am considering changing my major because of the class.

I originally was not thrilled at the prospect of having to do a community project, but now am so thankful that you made us do it!

SOURCE: Davies (2016).

working in 24 teams. Three teams planned rain gardens for an elementary school. They assessed the site and its drainage patterns and presented their designs at a meeting of teachers and the parent-teacher association. The students whose design was selected by the school adapted it in response to questions raised at the meeting, and in a subsequent semester, the students constructed the garden in collaboration with school volunteers (see Figure 4-4).

Students reflect on their experiences after the course, and give the service-learning component high ratings and positive comments, Gilbert reported (see Table 4-2). Many have later been hired by project clients or serve as mentors themselves. The stakeholder meetings are excellent opportunities for engagement between the university and the community, often bringing together groups who rarely collaborate. Many of the project designs have been implemented, he added.

Gilbert closed with an overview of some of the challenges he encountered. Coordinating approximately 140 students engaged in 35 new projects each year is difficult, he noted. Coming up with new projects each year can be challenging, and he said it is important to accommodate different sorts of experiences and to allow students to learn from their mistakes. It can be hard to be sure the project experiences strike a good balance between learning by doing and "catastrophic failure," he observed,

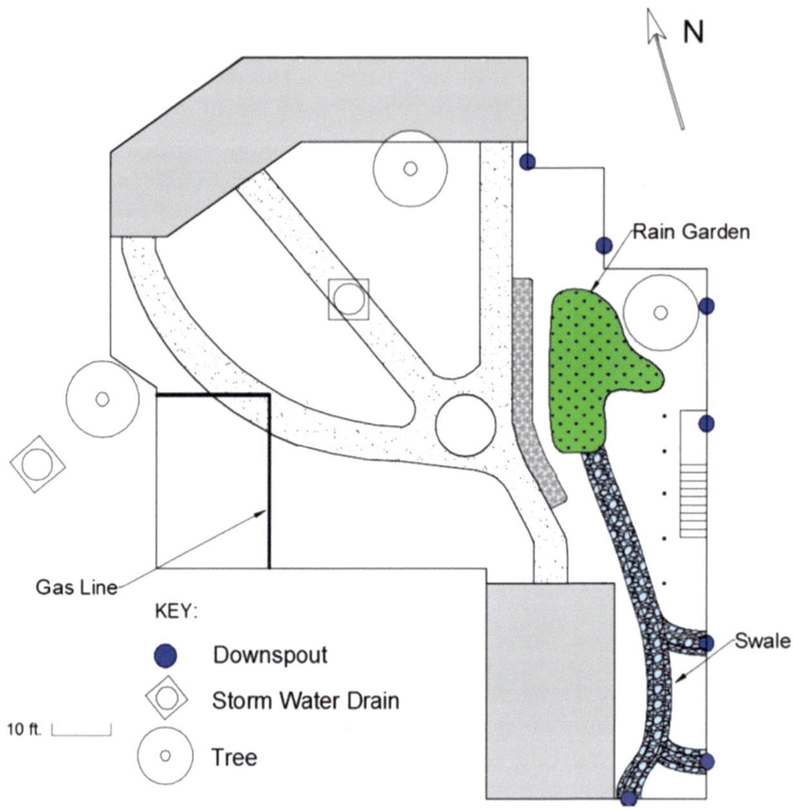


**FIGURE 4-4** Rain garden design developed through a service-learning project.
SOURCE: Gilbert (2016).

and to confirm that all the students benefit from their experiences. Gilbert is also mindful that he and the students relied heavily on the clients and mentors throughout the projects. It was easy for the workload to become unmanageable, both for faculty and students, he added.

### Lessons from These Examples

In follow-up discussion of these examples, Fortner summarized some benefits of service-learning experiences. These experiences give students the chance to develop new skills, as well as access to resources and opportunities not available in the classroom, she observed. They often integrate academic disciplines and people from many sectors and jobs in pursuit of a shared goal. This integration, in turn, benefits institutions' relation-

**TABLE 4-2** Sample Student Reflections after University of Texas Service-Learning Project

| What did you do well? | What was the outcome of what you did well? | What would you like to have changed that did not go so well? |
|---|---|---|
| We adapted to the change of our project. The elementary school is constantly working on our habitat area and building new projects, such as butterfly gardens, in the area. We found a way to work around that and keep much of the school's existing amenities. | We changed our location of the rain garden. This also changed our size, design, and swales. We were able to address the flooding problems the school has been having at the same time. | It took us a long time to decide what we wanted to do. By the time we changed the location of the rain garden, we had already done a lot of work for our previous location. We had to work a little more just to make this project work out. |
| We communicated the information in a clear and consistent manner. | The stakeholders seemed very excited about our idea and had relatively few concerns about the project. | I would have liked to have been better informed on the species of plants that we'd selected. I'm admittedly a little lacking in my knowledge of plant selection for gardens so answering questions about what other plants we might recommend proved to be difficult. So I guess a little more background research to better prepare for questions and not just giving the information that was planned for the presentation. |

SOURCE: Gilbert (2016).

ships with their communities and also opens students' worlds and their sense of the contributions they can make. Fortner also noted the many challenges in coordinating these activities, and participants shared their perspectives on those. The challenges include institutional limitations on flexibility that hamper service-learning projects, one noted. Another agreed, pointing out that institutional cultures are difficult to change and some institutions need encouragement to treat service-learning as a valuable part of the academic experience. In many cases, it falls to individual faculty members to make a pitch to a provost or university president, one participant pointed out, adding that it is especially effective to provide written documentation and data demonstrating outcomes and benefits to students. A participant also noted the difficulty of sustaining these programs and transferring their stewardship to new faculty. Coordinating and articulating procedures and expectations across classes and semesters is challenging, and establishing useful archives is "a nontrivial problem," this person noted.

## DISCUSSION: SIMILARITIES AND DIFFERENCES

In a second breakout discussion, participants met in small groups to discuss elements that are common across service-learning experiences and aspects that are unique to particular circumstances. The three groups were each asked to consider the issues from one perspective, that of:

- different geoscience disciplines, for example, oceanography or atmospheric science;
- different types of institutions, for example, community colleges, primarily undergraduate institutions, comprehensive universities, or research universities; or
- different types of students, for example, students at the honors or introductory level, members of historically underrepresented groups, local students or those attending school far from home, and first-generation college students.

### Looking Across Geoscience Disciplines

The group assigned to consider the perspectives of geoscience disciplines identified several elements that are shared:

- Service-learning projects provide an opportunity for people who may not have done so to reflect on the role of science in their environment or surroundings.
- Themes, such as hazards, resilience, sustainability, and economic issues, arise through community problems regardless of the discipline through which they are addressed. The interaction of built, natural, and human systems is likely to come up, as are the connections between public health and the health of ecosystems. Environmental justice is another shared theme.
- The explicit focus on interdisciplinary collaboration that is frequently an element of service-learning projects is valuable for every discipline.

- The full engagement of community members of all ages and roles is an asset to service-learning projects, regardless of discipline.
- Safety issues—related to work in the field or in the laboratory or to interactions between participants—are important regardless of discipline.

The group also identified a few features that are unique to particular geoscience fields. They observed that the courses that are most easily adapted to include service-learning projects tend to rely heavily on measurement and to focus on communication, and to be ones that already incorporate material from more than one discipline. Other points concerned specific topics likely to be suitable for service-learning projects. For example, the private sector is in need of weather forecasts with immediate economic use and impact, as well as information about short-term hazards, which are in the purview of the atmospheric sciences. Communities' needs will vary depending on their geographical location, but those that rely economically on the ocean ecosystem, for example, may have great need for analysis and other services that students of oceanography could provide. Students of geology and geophysics may provide valuable insights about long-term hazards, and students of environmental science, geophysics, or hydrology can assist with monitoring or mapping of pollutants.

This group also noted that many of the good ideas for structuring group service-learning experiences apply across disciplines. Regardless of discipline, it is important to incorporate and honor the knowledge the community members bring, and also to think carefully about the long-term implications for the community of the findings and conclusions the service-learning project yields. Ethical considerations the project may raise deserve careful consideration. A concern that may not always be obvious is that the students involved in the project should be allies, not competitors, with professionals who could do the work.

It is important for faculty to consider the skills and experiences students are likely to need in the workplace as they match students to projects, and also to be aware of safety, risk management, and liability concerns. Regardless of the discipline, it is also important to have a well-planned communications strategy, and faculty in all disciplines benefit from institutional support in planning and overseeing successful projects.

## Looking Across Types of Institutions

Faculty at all types of institutions have many of the same reasons for including service-learning in their courses, the second group found. Faculty members want to provide engaging learning experiences that will help to recruit and retain students to their disciplines and that will provide value to the community. They also share many of the same goals for the students who participate in these experiences: they gain communication skills, civic awareness, and the capacity to use what they have learned in the classroom to bring about positive changes. They use these experiences in part to help students to develop a connection with the communities in which the project is done. Faculty who offer service-learning experiences also go about it in much the same way, regardless of institutional type. They help their students plan and prepare, support them as they implement their plans and synthesize what they have learned, and guide them in reflecting on the experience. Faculty at any type of institution benefit from having support from that institution, yet shortages of resources—time, money, and space—needed for successful service-learning projects are issues at all types of institutions, the second group found.

On the other hand, the group observed that service-learning projects need to be designed to suit the needs and constraints of the institution in which they take place. Institutions serve different types of students so there is variation in the kind of involvement students have time for outside of class, how easily and far they might be able to travel, and the goals they have for their undergraduate years. Institutions vary also in their priorities and in their relationships with the communities in which they are situated. Those that serve primarily local students, for example, may have a more pronounced focus on their immediate surroundings than those that serve a more geographically diverse student body. Institutions have different structures for supporting activities such as service-learning and use different terminology to talk about programs of this nature (e.g., experiential learning, civic engagement). Some have centers dedicated to facilitating such experiences, but institutions vary significantly in their resources—in terms of finances, faculty time, availability of graduate students, and other elements that contribute to successful service-learning programs.

The group suggested that service-learning programs be structured in a way that suits the number of students involved and the level of support available from mentors and the administration. Many students at community colleges, the group noted, work full time and simply are not able to commit significant time to a project that requires travel, but they can still benefit from one that permits much of the work to be done on campus or independently. Scaffolding the elements of the service-learning experience (e.g., development of proposal, progress updates, collation of data) so it is not overwhelming for students is likely to be helpful regardless of institution type, and grade weights should reflect the nature of the commitment being asked of students. Regardless of its type, any institution would benefit from developing long-term relationships with community partners that can be the basis for different types of projects across courses and disciplines, the group suggested.

### Looking Across Types of Students

All students face challenges in dealing with the unpredictable nature of real-world service-learning projects, the third group found. Students who perform well in these projects are often not those who perform at the highest levels academically, the group noted, but those who bring other valuable skills to the experience. Regardless of the kinds of students involved, faculty need wisdom and judgment as they guide students, steering them toward collaboration with community partners (as opposed to volunteering to help those less fortunate), and also knowing when to step back and when to intervene.

The third group's suggestions for designing programs reflected the view that guiding students also requires understanding of differences among students. Faculty should plan in advance how they will guide students to value the culture, values, and ways of knowing they will encounter as they work with their community partners. At the same time, the group reported, it is important to focus on areas where partnership is requested from the community, as opposed to imposing ideas on community partners. The group advised that faculty set clear expectations for students about following norms and customs they encounter, and be explicit in encouraging students to be sensitive and open to new ways of thinking. The involvement of peers as well as seasoned scientists can be very helpful in providing credible guidance and perspective to undergraduates, but mentoring should be flexible and responsive to the strengths and traits of the students involved and the needs of the community.

# 5

# Looking Forward

The final workshop session was an opportunity for reflection on the future of geoscience service-learning and the primary messages from the presentations and discussions. Committee chair Cathy Manduca offered her thoughts on the answers offered by workshop participants to the guiding questions for the workshop. Committee member Susan Sullivan suggested ways that future research could support the growth and improvement of geoscience service-learning programs, and committee members each identified their primary take-away messages from the workshop.

## REFLECTIONS ON THE GUIDING QUESTIONS

Manduca offered her own impressions of the key ideas presented in answer to the guiding questions for the workshop (see Chapter 1). One question was about how geoscience service-learning should be defined. Fundamentally, Manduca suggested, service-learning is an experience in which students partner with others to address issues of communal importance using skills and knowledge that are integral to the learning objectives of the geoscience course. Yet, she added, these kinds of projects are done in many contexts, with students at different levels, in different kinds of institutions, in all geoscience disciplines, and at a wide variety of scales of implementation.

Service-learning programs have numerous benefits, Manduca went on. They may support students' intellectual development by engaging them in higher-order thinking and helping them develop the desire and the capacity for lifelong learning. Surveys of students involved in these programs and the faculty who work with them indicate that the students are highly engaged, and there is reason to believe that they learn more and at a higher level than they might without that component, she commented. By involving students in situations in which they can have an impact on real-world problems and interact with adults in a variety of roles, these experiences may build students' confidence in what they as individuals can accomplish. They may also help students to explore ideas for their future careers, develop skills that will be attractive to employers and a portfolio of experiences that demonstrate those skills, and expose them to resources and opportunities beyond their institutions.

These experiences also may bring benefits to the institutions, Manduca noted. For institutions that include service to the community or society or the development of citizenship in their explicit missions—as many historically black colleges and universities and those with a focus on serving Native American students do—service-learning is an educational experience that actively promotes that mission. The intellectual development these experiences promote is a benefit not only to students but also to their institutions. Successful service-learning projects can enhance an institution's visibility and reputation, and also make it easier for it to recruit and retain students. Service-learning projects can also enhance the institution's reputation by demonstrating the expertise of its faculty and students and by making that expertise available to meet societal needs. These projects can also help develop the institution's relationship with its community and open up new resources for its students and faculty. Faculty members, in turn, have the chance both to learn and to have their own opportunities to make a contribution. The question of benefits to faculty generated some discussion, Manduca noted. The burden on them can be high, and the degree of support and recognition they receive for this work varies significantly across institutions. She suggested that many faculty members who use service-learning in their courses report finding the experience extremely rewarding but also have concerns about how to use service-learning most effectively in their teaching.

In terms of best practices, Manduca noted, there is relatively little documentation. Published papers and examples provide some insights, and handbooks and toolkits also provide guidance and ideas (see Appendix C for examples). However, because service-learning is very context-specific, any examples and ideas will need to be adapted to circumstances, she emphasized, and used as the basis for innovation in a local context. One strategy that may be valuable in any context, she noted, is to structure and explain the tasks and stages necessary to successfully complete the project, so that students fully understand the expectations and can reap benefits from each aspect, even if the project is not 100 percent successful. Mentoring, opportunities for reflection, and assessment and evaluation also seem likely to be beneficial regardless of the context, Manduca added.

Other less concrete factors are no less important in the design of service-learning projects, Manduca observed. The selection of projects that are suitable for partnerships between students and communities requires careful thought. As the projects take shape, faculty should guide students in thinking about both intended and unintended consequences, taking into account practical and ethical issues, she observed. The projects should be opportunities for true partnership, where the need being met is one that is of genuine importance to the community or group, and that will make use of the expertise of students and community partners, she added. Faculty should be ready to offer guidance as the interactions within student groups and among students and community partners develop, with an eye to helping group members appreciate the sensitivities of others.

Service-learning experiences are intended to draw students, and especially those from historically underrepresented populations, into geoscience disciplines. Manduca commented that service-learning has the potential to meet those goals by demonstrating the contributions the geoscience disciplines can make to solving community challenges, and the professional opportunities open to students who choose geoscience majors. The opportunity to engage in interdisciplinary problem solving can be a draw for students—as can the opportunity to address issues they care deeply about—particularly those associated with protecting the environment. However, there is little research to document the effects of these opportunities on efforts to recruit students from historically underrepresented groups into geoscience majors. Peer mentors and role models are likely to help draw students in and support them, she noted. Attention to obstacles that may deter some students, and to the nature of the experiences they have, can help strengthen the appeal of geosciences among these groups, she added.

There are also many reasons to believe that service-learning builds skills that are needed and valued by employers, Manduca said. Problem solving, communication, and teamwork are among the skills employers often prize that are developed through service-learning. She noted that service-learning experiences seem to contribute to success in students' undergraduate studies and in their first postcollege employment experiences (see Chapter 3). Service-learning is also an opportunity for students to explore potential careers and develop portfolios they can use to demonstrate their expertise to potential employers.

In closing, Manduca noted that the workshop presenters and participants had identified many obstacles to implementing service-learning programs and expanding access to successful ones. She noted that the importance of support from both the institution and the academic department was highlighted throughout the workshop as a key to helping faculty and students succeed in these experiences.

## SUGGESTIONS FOR FUTURE RESEARCH

The body of research that is relevant to service-learning in the geosciences is scattered and incomplete, Sullivan commented, so the need is great. She offered her priorities for research, structured around the benefits service-learning can have for addressing complex problems and for benefiting students, communities, and the faculty and institutions who support them.

To promote and support the kinds of projects that address challenging and important social issues, Sullivan suggested, it would be helpful to learn more about how to support faculty in taking on problems that are of importance to communities and draw on multiple disciplines. Given that students are still developing their mastery of relevant skills and knowledge, she added, it would also be useful to learn more about the kinds of contributions they can realistically be expected to make. Identifying service-learning goals that serve both students' and community partners' needs, while also addressing complex societal problems is another area in which further research could be helpful.

Sullivan also suggested specific research questions related to benefits for students, faculty and institutions, and communities:

*Questions Related to Benefits for Students*

- What are the benefits and costs of service-learning projects for different student groups, including historically underrepresented groups?
- How do precollege experiences support or hinder undergraduate service-learning readiness?
- How does service-learning affect students' workforce experiences?
- How does service-learning deepen student outcomes in the geosciences?
- Are there ways to match particular geoscience students and types of experiences that maximize the benefits of a service-learning project?
- How does student engagement change?
- What service-learning outcomes are valued by employers?
- How might service-learning increase diversity in the geoscience workforce?

- What geoscience workforce skills may be deepened through service-learning?
- How does service-learning experience affect students' later experiences in the workforce?

*Questions Related to Benefits for Faculty and Institutions*

- What support, training, and rewards systems are most likely to yield programs that demonstrate best practices?
- What are the costs and benefits of service-learning programs for faculty and institutions?
- How can faculty use service-learning activities to deepen rather than distract from course learning?
- How do faculty members' research agendas change as a result of their service-learning programs?
- How do best practices vary by type of college or university?

*Questions Related to Benefits for Communities*

- What types of opportunities or challenges are associated with different kinds of community projects?
- How should all participants be prepared to develop effective partnerships?
- What types of community-based projects are suitable for the geosciences?
- How do previous experiences with the geosciences or with educational institutions influence community expectations for service-learning projects?

Linda Silka (see also Chapter 2) volunteered additional ideas and research questions, based partly on the workshop discussions and partly on her own observations. She suggested several types of analyses that would be valuable in the context of geoscience service-learning (see Box 5-1). Silka also suggested some questions about geosciences service-learning that researchers might explore.

*How are service-learning partnerships initiated?* Research on partnerships between communities and universities has focused on the logistical, ethical, and power issues that influence the way such partnerships are initiated. Additional research in this area could help support successful partnerships.

*Which faculty members might be well suited to participate in service-learning?* There are a number of possible reasons why some faculty members are more interested in or have more successful experiences with

**BOX 5-1**
**Research Topics Suggested by Linda Silka**

- Analysis of Institutional Support Structures: Examination of current support structures (or lack thereof) in different types of campuses and how those structures provide support, such as pre-established relationships with community partners, assistance with transporting students to service-learning sites off campus, and guidance in how to assess service-learning projects.
- Analysis of a Program's Theory of Change: Investigation of how geoscientists conceptualize what is needed for service-learning to be effective in meeting their objectives. This type of analysis, which is often part of program evaluation, can be helpful in identifying possible mismatches between the outcomes expected and the activities carried out.
- Effort Analysis: A systematic assessment of the effort required by a project and whether that level of effort is commensurate with the expected outcome and likely impact.
- Research on Boundary: Spanning. Some research indicates that organizations that have developed the practice of reaching across boundaries or disciplines and incorporating ideas and practices from beyond their own circle, often called boundary-spanners, can contribute to the success of projects involving diverse groups. Further study of boundary-spanning organizations and individuals would be very useful for geoscience service-learning.
- Social Network Analysis: Analysis of data on the contacts and working relationships involved in a project or other endeavor is often done with large-scale interdisciplinary initiatives and has been effective in uncovering ways in which different groups or subgroups are in communication or isolated, and to identify communication gaps or opportunities.
- Partner Preferences: Researchers in other higher education contexts have begun to study how different factors and contexts may be predictors of potential partners' interest in working together—insights that could be helpful in setting up productive service-learning partnerships.

SOURCE: Silka (2016b).

service-learning, Silka noted. Some have suggested that tolerance for ambiguity is an important asset for faculty members undertaking a complex, multiparty, multidisciplinary initiative such as service-learning, but additional research would be helpful, in her view.

*Where do service-learning projects tend to take place?* A service-learning project that can be carried out near campus is easiest for students and faculty but projects that require a modest amount of travel can work well. Alternatively, remote participation is a possibility, though some approaches might only work face to face. Better understanding of these factors could be very helpful to faculty and administrators, Silka noted.

*How can service-learning projects be responsive to "abrupt problems"?* Often, new problems or issues come into the public eye unexpectedly, Silka commented. Events such as earthquakes, hurricanes, flooding, or water quality problems such as those in Flint, Michigan, might be valuable service-learning opportunities, Silka commented. However, setting up service-learning projects often takes considerable advance planning, and she suggested that sharing of information about ways to structure service-learning programs to be flexible enough to accommodate last-minute needs could be very useful.

## CLOSING THOUGHTS

The workshop ended with closing thoughts from committee members and from Amanda Adams of the National Science Foundation (NSF), the workshop sponsor. The committee members were each asked to identify a single point brought out in the workshop that he or she regarded as particularly important.[1]

Ed Laine identified the importance of training faculty members to do service-learning as key. This can be done through external institutions such as Campus Compact,[2] he noted, or through mentoring by colleagues in the geosciences or other departments within the university. Colleagues from beyond the department and the discipline can be extremely helpful, he emphasized, as many aspects of the experience are relevant across content areas. "People just need access to ideas that work," he noted, and he suggested that a Website or other clearinghouse—not for best practices, since needs and circumstances vary so much, but for ideas and methods that are applicable across fields—would be very useful.

Sue Ebanks also emphasized that service-learning is "not a one size fits all." There are not models that will fit all circumstances and each experience needs to be customized. The process is "messy and dynamic," she added and all participants, including faculty and graduate student mentors, need to be flexible and "ready for anything" if the project is to be a successful experience for all. Programs that are set up this way, she suggested, build community trust even when outcomes are not exactly as expected. If students believe their teachers and mentors "are with them whether it goes well or not," they will learn more from the experience. This also should be the goal for assessment and evaluation—to learn from experience, she commented.

For Sarah Fortner, the most important point was that the geosciences

---

[1]Not all of the committee members were present for this final discussion.

[2]A coalition of colleges and universities focused on civic engagement discussed in Chapter 3; see http://compact.org/who-we-are/ [September 2016].

teach skills and ideas that are ideal for addressing "wicked problems," and service-learning experiences are opportunities to directly engage those ideas and skills. Thus, she explained, the importance of the issues these projects address helps to expand the diversity of the audience that is engaged with the geoscience fields.

Art Goldstein pointed out that many workshop speakers stressed that administrative support for service-learning projects is vital, but that "we haven't been very specific." He said he hopes the field can provide clear guidance about what support is needed and how it will help expand service-learning. Clarity regarding the roles and responsibilities of students, faculty members, the administration, and the host site, he noted, would be valuable.

Susan Sullivan noted that geoscience service-learning has the potential to be "transformative" for individuals and communities, but that more work is needed on how to use assessments, course evaluations, and research to improve and expand access to these experiences.

In discussion, participants offered additional thoughts. One expressed the hope that geoscience service-learning would play a key role as society addresses "the grand challenge of sustainability." Another commented that service-learning is key to the future of the geosciences, noting that enrollment has already expanded significantly and that service-learning has "a ripple effect" that can reach broad audiences and engage them in protecting the earth. A final thought from a participant was that "service-learning is a hidden gem—there is so much interesting stuff going on that broader communities would want to know about."

The workshop closed with thanks from Adams. Interest at NSF in service-learning grew out of its commitment to both increasing science literacy and promoting community-relevant science, she explained. NSF staff hoped the workshop would identify answers to their questions about the nature, extent, and contributions of service-learning in the geosciences and also identify key questions that still need to be answered. She expressed her appreciation to the committee for a productive set of presentations and discussions.

# References

Ash, S.L., and Clayton, P.H. (2009). Generating, deepening, and documenting learning: The power of critical reflection in applied learning. *Journal of Applied Learning in Higher Education, 1,* 25-48.

Astin, A.W., Vogelgesang, L.J., Ikeda, E.K., and Yee, J.A. (2000). *How Service Learning Affects Students.* Paper 144. Los Angeles: Higher Education Research Institute, University of California. Available: http://heri.ucla.edu/PDFs/HSLAS/HSLAS.PDF [December 2016].

Battistoni, R.M., and Longo, N.V. (2006). *Connecting Workforce Development and Civic Engagement: Higher Education as Public Good and Private Gain.* Danvers, MA: North Shore Community College Public Policy Institute. Also available: http://www.learningace.com/doc/2090672/9d8152ff2d44144b29b7271dd7e37e4b/connecting-workforce-development-civic-engagement [January 2017].

Billig, S.H. (2009). Does quality really matter? In B.E. Moely, S.H. Billing, and B.A. Holland (Eds.), *Creating Our Identities in Service-Learning and Community Engagement: Advances in Service-Learning Research* (pp. 131-157). Scottsdale, AZ: Information Age.

Cajete, G. (1999). *Igniting the Sparkle: An Indigenous Science Education Model.* Skyland, NC: Kivaki Press.

Celio, C.I., Durlak, J., and Dymnicki, A. (2011). A meta-analysis of the impact of service-learning on students. *Journal of Experiential Education, 34*(2), 164-181.

Conway, J.M., Amel, E.L., and Gerwien, D.P. (2009). Teaching and learning in the social context: A meta-analysis of service learning's effects on academic, personal, social, and citizenship outcomes. *Teaching of Psychology, 36*(4), 233-245.

Davies, C. (2016). *Service-Learning in Undergraduate Geosciences.* PowerPoint presentation for the Workshop on Service-Learning in Undergraduate Geosciences, April 20-21, 2016, National Academies of Sciences, Engineering, and Medicine, Washington, DC. Available: http://sites.nationalacademies.org/cs/groups/dbassesite/documents/webpage/dbasse_172174.pdf [December 2016].

Eyler, J.S., Giles Jr., D.E., Stenson, C.M., and Gray, C.J. (2001). *At A Glance: What We Know about the Effects of Service-Learning on College Students, Faculty, Institutions, and Communities, 1993-2000* (third edition). Available: http://www.nationalservice.gov/volunteering-pathway-employment-report [June 2016].

Gilbert, R.B. (2016). *Engineering Professionalism Objectives.* PowerPoint presentation for the Workshop on Service-Learning in Undergraduate Geosciences, April 20-21, 2016, National Academies of Sciences, Engineering, and Medicine, Washington, DC.

Hoisch, T.D., and Bowie, J.L. (2010). Assessing factors that influence the recruitment of majors from introductory geology classes at Northern Arizona University. *Journal of Geoscience Education, 58,* 166-176.

Huntoon, J.E., and Lane, M.J. (2007). Diversity in the geosciences and successful strategies for increasing diversity. *Journal of Geoscience Education, 55,* 447-457.

Kober, N. (2015). *Reaching Students: What Research Says about Effective Instruction in Undergraduate Science and Engineering.* Board on Science Education, Division of Behavioral and Social Sciences and Education. Washington, DC: The National Academies Press.

Levine, R.E., Gonzalez, R., Cole, S., Fuhrman, M., and Le Floch, K.C. (2007). The geoscience pipeline: A conceptual framework. *Journal of Geoscience Education, 55,* 458-468.

National Science Foundation. (2012). Science and Engineering Indicators. *Completions Survey: 2002-2012.* Available: https://www.nsf.gov/statistics/2015/nsf15321/pdf/tab10.pdf [June 2016].

National Science Foundation and National Center for Science and Engineering Statistics. (2015). *Women, Minorities, and Persons with Disabilities in Science and Engineering: 2015.* Special Report NSF 15-311. Arlington, VA. Available: https://www.nsf.gov/statistics/2015/nsf15311/start.cfm [December 2016].

O'Connell, S. (2016). *Diversity and Service Learning in the Geosciences: Opportunities for Exploration.* Paper commissioned for the Workshop on Service-Learning in Undergraduate Geosciences, April 20-21, 2016, National Academies of Sciences, Engineering, and Medicine, Washington, DC. Available: http://sites.nationalacademies.org/DBASSE/BOSE/CurrentProjects/DBASSE_169060 [October 2016].

Orr, D.W. (2004). *Earth in Mind.* Washington, DC: Island Press.

Savanick Hansen, S., and Fortner, S. (2016). *Geoscience Service-Learning Literature Themes.* Paper commissioned for the Workshop on Service-Learning in Undergraduate Geosciences, April 20-21, 2016, National Academies of Sciences, Engineering, and Medicine, Washington, DC. Available: http://sites.nationalacademies.org/cs/groups/dbassesite/documents/webpage/dbasse_171831.pdf [October 2016].

Silka, L. (2016a). *Service Learning in the Geosciences: Opportunities for Innovation.* Paper commissioned for the Workshop on Service-Learning in Undergraduate Geosciences, April 20-21, 2016, National Academies of Sciences, Engineering, and Medicine,Washington, DC. Available: http://sites.nationalacademies.org/cs/groups/dbassesite/documents/webpage/dbasse_171830.pdf [October, 2016].

Silka, L. (2016b). *Service Learning in the Geosciences: Opportunities for Innovation.* PowerPoint presentation for the Workshop on Service-Learning in Undergraduate Geosciences, April 20-21, 2016, National Academies of Sciences, Engineering, and Medicine, Washington, DC. Available: http://sites.nationalacademies.org/cs/groups/dbassesite/documents/webpage/dbasse_172170.pdf [December 2016].

Stokes, P.J., Levine, R., and Flessa, K.W. (2015). Choosing the geoscience major: Important factors, race/ethnicity, and gender. *Journal of Geoscience Education, 63,* 250-263.

Szymanski, D.W. (2016). *Why Is Service-learning Important in the Geosciences?* PowerPoint presentation for the Workshop on Service-Learning in Undergraduate Geosciences, April 20-21, 2016, National Academies of Sciences, Engineering, and Medicine, Washington, DC. Available: http://sites.nationalacademies.org/cs/groups/dbassesite/documents/webpage/dbasse_172168.pdf [December 2016].

Velasco, A.A., and Velasco, E.J. (2010). Striving to diversify the geosciences workforce. *EOS, 91*, 289-296.

Whitney, D., Behl, R., Ambos, E., Francis, R.D., Holk, G., Larson, D., Lee, C., Rodrique, C., and Wechsler, S. (2005). Ethnic differences in geoscience attitudes of college students. *EOS, 86*(30), 277-279.

Youth Rebuilding New Orleans. (2015). *Service Learning*. New Orleans, LA: Author. Available: http://yrno.com/volunteer-work-new-orleans/service-learning/ [December 2016].

# Appendix A

# Workshop Agenda and Participants List

**Service-Learning in Undergraduate Geosciences Workshop**
**April 20-21, 2016**
**National Academy of Sciences Building, Room 125**
**2101 Constitution Ave, NW, Washington DC**

*April 20*

| | |
|---|---|
| 8:30 a.m. | Breakfast and Networking |
| 9:00 a.m. | **Welcome and Workshop goals**<br>**Cathy Manduca, Steering Committee Chair** |
| 9:15 a.m. | Why is service-learning important in the geosciences?<br>**David Szymanski, Bentley University** |
| 9:45 a.m. | Q&A |

**Session I: Understanding the Breadth of Service-Learning Implementation Across Disciplines**

Moderator: Ed Laine, Steering Committee Member

| | |
|---|---|
| 10:00 a.m. | The Promise of Service-Learning: Outcomes and Challenges<br>**Linda Silka, University of Maine** |
| 10:30 a.m. | Q&A |
| 10:45 a.m. | Break |

11:00 a.m. Reflect on Learning Goals: Breakout Session

What have been your experiences/observations about service-learning? What is valuable about service-learning to your students, faculty, institution, and community? What has gone well in your service-learning experience? What needs to be improved? (30 minutes)

What are the barriers to the adoption of service-learning within the geoscience profession? What barriers are there to adoption of service-learning at your institution? Brainstorm strategies for overcoming these barriers. (30 minutes)

***Each group should prepare slides with bullet points***

- Slide #1-What is being done well?
- Slide #2-What areas need attention?

12:00 p.m. Reporting back from breakout sessions

- Moderated by Ed Laine
  - *First group will present its top three points from each slide*
  - *Subsequent groups will attempt to give three new points each*

12:30 p.m. *Lunch*

**Session II: Understanding What Is Known about the Impact of Service-Learning Experiences on Students from Historically Underrepresented Populations**

Moderator: Sue Ebanks, Steering Committee Member

1:30 p.m. Participation in geosciences and service-learning by historically underrepresented populations
**Suzanne O'Connell, Wesleyan University**

1:50 p.m. Panel of commenters on what the presentation suggests about impact (positive, negative, and unique opportunities)

- **Antony Berthelote, Salish Kootenai College (remote)**
- **Ben Cuker, Hampton University**

2:10 p.m. Q&A

2:30 p.m. Break

**Session III: Understanding How the Geoscience Community Defines and Implements Service-Learning**

Moderator: Sarah Fortner, Steering Committee Member

2:45 p.m. Categorizing and synthesizing information about service-learning experiences in the geosciences
**Suzanne Savanick Hansen, Macalester College**

3:05 p.m. Specific examples of service-learning
- **Caroline Davies, University of Missouri, Kansas City**
- **Bob Gilbert, University of Texas, Austin (remote)**

3:25 p.m. Q&A

3:45 p.m. Breakout Session: Reflect on the common elements of service-learning in the geosciences and consider which aspects are unique to the various groups within your assignment.

- *Group #1-Disciplines of the Geosciences (NAS 125)*
  - *Oceanography, atmospheric science, etc.*
  - *Remote participants may only participate in this breakout group*
- *Group #2-Institution Types (NAS 227)*
  - *Community colleges, primarily undergraduate institutions, comprehensive universities, research universities, etc.*
- *Group #3-Types of Students (NAS 360)*
  - *Honors, introductory, historically underrepresented, local, attending school far from home, first-generation college students, etc.*

***Each group should prepare slides with bullet points***
- o Slide #1-common elements
- o Slide #2-unique features
- o Slide #3-relevant design principles for service-learning in the context of your group's assigned topic

4:45 p.m. Reporting back from the breakout session
- Moderated by Sarah Fortner
  - o *First group will have 5 minutes to present its top three points*
  - o *Subsequent groups will attempt to give three new points each in their 5-minute slots*

Wrap Up

5:15 p.m. Review of the day
**Cathy Manduca**

5:30 p.m. Adjourn

6:00 p.m. Dinner on your own

*April 21*

8:30 a.m. Breakfast and Networking

9:00 a.m. Recap of Day #1 and overview of Day #2 Plans
**Cathy Manduca**

### Session IV: Understanding How Service-Learning Experiences Could Enhance The Competitiveness of Geoscience Students in the Workforce

Moderator: Art Goldstein, Steering Committee Member

Background Material:
*Geoscience Employers Workshop Summary* (May 2015) Available: http://www.jsg.utexas.edu/events/files/Summit_Webnar_2015.pdf and http://www.jsg.utexas.edu/events/files/Employers_Workshop_outcomes.pdf [January 2017]

*Connecting Workforce Development and Civic Engagement: Higher Education as Public Good and Private Gain*
Available: http://www.tncampuscompact.org/files/wfd.pdf [January 2017]

9:20 a.m. Panel discussion on service-learning and workforce

1. Is there evidence that service-learning helps students get jobs?
2. Does service-learning help create a sense of professionalism among students (such as in the way they write, speak, and/or present themselves)?
3. Is there any evidence that service-learning helps create a more positive impression of a college or university among townspeople?
4. What can be done to create more service-learning opportunities?
    - **John Gierke**, Michigan Technological University and Peace Corps Master's International Program
    - **Amy Cohen**, George Washington University
    - **Eric Riggs,** Texas A&M

10:20 a.m. Q&A

10:45 a.m. Break

11:00 a.m. Breakout Session: Planning service-learning projects in the geosciences to enhance workforce competitiveness

- How can service learning projects be structured to better prepare students for employment?
- What type of faculty oversight and site oversight is needed for good service-learning projects? What are the various roles of faculty, administrators, and site supervisors in service-learning project design and implementation?
- Does service-learning have a role helping students learn 21st century skills (communication, teamwork, etc.)?
- Are certain geoscience sectors or employers more amenable to service-learning projects?
- What role should funders play in expanding service-learning opportunities for geoscience students?

***Each group should prepare slides with bullet points***

12:00 p.m. Lunch

1:00 p.m. Reporting back from breakout sessions

1:45 p.m. Discussion of Potential Research Questions
**Susan Sullivan, Steering Committee Member**
- Comments from the audience

2:15 p.m. The future of service-learning in the geosciences
**Cathy Manduca**
- Committee members and participants share the messages they heard during the workshop

3:00 pm. Adjourn

## PARTICIPANT LIST

| Name | Affiliation | Attending |
|---|---|---|
| Cathy Manduca | Carleton College | In Person |
| Antony Bethelote | Salish Kootenai College | Remote |
| Sue Ebanks | Savannah State University | In Person |
| Sarah Fortner | Wittenberg University | In Person |
| Art Goldstein | Bridgewater State College | In Person |
| Ed Laine | Bowdoin University | In Person |
| Susan Sullivan | University of Colorado Boulder | In Person |
| David Szymanski | Bentley University | In Person |
| Linda Silka | University of Maine | In Person |
| Suzanne O'Connell | Wesleyan University | In Person |
| Benjamin Cuker | Hampton University | In Person |
| Suzanne Savanick Hansen | Macalester College | In Person |
| Caroline Davies | University of Missouri, Kansas City | In Person |
| Bob Gilbert | University of Texas, Austin | Remote |
| John Gierke | Michigan Technological University | In Person |
| Amy Cohen | George Washington University | In Person |
| Eric Riggs | Texas A&M University | In Person |
| Philip Garwood | Cape Fear Community College | In Person |
| Laura Lukes | George Mason University | In Person |
| Amanda Adams | National Science Foundation | In Person |
| Karen Berger | University of Rochester | In Person |
| Laura Guertin | Pennsylvania State University, Brandywine | In Person |
| Carolyn Wilson | American Geosciences Institute | In Person |
| Jennifer Nelson | Indiana University–Purdue University Indianapolis | In Person |
| Heather Houlton | American Geosciences Institute | In Person |
| Alexandra Moore | Cornell University | In Person |
| Pranoti Asher | American Geophysical Union | Remote |
| Jennifer Silko | Pennsylvania State University, Harrisburg | In Person |
| Jill Singer | SUNY-Buffalo State | In Person |
| Ann Maest | Buka Environmental | Remote |
| Livia Lam | Learning Policy Institute | Remote |
| Adriandinny Braz | N/A | Remote |
| Jennifer Albert | The Citadel | Remote |
| Alexandra Davatzes | Temple University | Remote |
| Rick Olsen | USACE | Remote |

| Name | Affiliation | Attending |
|---|---|---|
| Ziya Ã–NCÃœ (Onis) | Suleyman Demirel University, Turkey | Remote |
| Sarah Truebe | Stanford University | Remote |
| Peter Wampler | Grand Valley State University | Remote |
| Daren Nelson | University of North Carolina, Pembroke | Remote |
| Elizabeth Dahl | Loyola University | Remote |
| Michael Guidry | University of Hawaii, Manoa | Remote |
| Wing Cheung | Palomar College | Remote |
| Deborah Anderson | St. Norbert College | Remote |
| Anne Hall | Emory University | Remote |
| Sadredin Moosavi | Gustavus Adolphus College/RCTC | Remote |
| Kaye Savage | Wofford College | Remote |
| Dana Vukajlovich | Bellevue College | Remote |
| Wendi Williams | Northwest Arkansas Community College and University of Arkansas at Little Rock | Remote |
| Amber Kumpf | Muskegon Community College | Remote |
| Timothy Kerr | Community College of Baltimore County | Remote |

# Appendix B

# Biographical Sketches of Committee Members and Presenters

**Antony Berthelote** *(committee member)* is the head of the Department of Hydrology at Salish Kootenai College. He was the lead groundwater researcher for the Environmental Protection Agency on the Milltown Dam Removal CERCLA (Superfund) Project. His specific research interests focus on fluid dynamics, computer modeling, groundwater-surface water interactions, protecting groundwater supplies, river/floodplain restoration, and shallow subsurface geophysics. He is the lead instructor in hydrology at Salish Kootenai College. As the college's American Indian Science and Engineering Society chapter advisor, he often mentors student community support activities and oversees student interns. He currently serves as cochair of the Geoscience Alliance and has served as an education and outreach committee member for the Consortium of Universities for the Advancement of Hydrological Sciences, Inc. since 2010. He holds a B.S. in geology from the University of Montana, an M.S. in geophysics from the University of Alaska, Fairbanks, and a Ph.D. in geosciences from the University of Montana.

**Amy Cohen** *(presenter)* is the executive director of the Honey W. Nashman Center for Civic Engagement and Public Service at George Washington University. Previously, she served as the director of Learn and Serve America, the federal service-learning program at the Corporation for National and Community Service, which is also the parent agency for AmeriCorps and Senior Corps. She has served as associate vice president for U.S. programs at Save the Children. She also worked at the University of Pennsylvania Netter Center for Community Partnerships, where she was an international leader in university-community partnerships.

**Benjamin Cuker** *(presenter)* is professor of marine and environmental studies at Hampton University. He created the American Society of Limnology and Oceanography Multicultural Program, the Multicultural Students at Sea Together, and the Hall-Bonner Program for Minority Doctoral Scholars in the Ocean Sciences. His current research includes limnology of turbid lakes, estuarine ecology, and hypoxia. He studies the depletion of oxygen in the Chesapeake Bay. He also studies the interaction between suspended sediments and nutrients in controlling community structure, general limnology, estuarine ecology, benthic ecology, and evolutionary ecology of aquatic communities. He received his Ph.D. in zoology with a minor in ecology from North Carolina State University.

**Sue Ebanks** *(committee member)* is assistant professor of marine and environmental sciences at Savannah State University. Her overall area of scientific research is the study of aquatic physiology and toxicology with particular focus on macroinvertebrates. She studies the effects of physically weathered and chemically dispersed crude oil on development and physiological processes in brine shrimp and marsh shrimp species. She is active in engaging students and underrepresented community youth in the exploration and monitoring of the Lower Ogeechee River Estuary. She actively advises the Savannah State Environmental Science Club and serves on the public policy committee of the National Association of Marine Laboratories. She has a B.S. and an M.S. from Savannah State University and a Ph.D. from the University of Miami.

**Sarah Fortner** *(committee member)* is assistant professor of geology and environmental science at Wittenberg University. Her teaching interests include inquiry-based teaching, climate literacy, and community engagement. Her goal is to improve student and research understanding of critical biogeochemical interactions at the Earth's surface. Students in her classes are involved in K-12 outreach at COSI (science museum) and work on community environmental issues. Presently, she is creating an agricultural sustainability curriculum through a national-level grant (National Science Foundation InTeGrate) that will be widely disseminated to other earth and environmental scientists. Additionally, she is investigating agricultural soil-water sustainability with the Agroecosystem Critical Zone Research Cluster at Ohio State University. She has a B.S. in geology from the University of Wisconsin–Madison and an M.S. and Ph.D. in geological and earth sciences from Ohio State University.

**John Gierke** *(presenter)* is a professor and chair of the Geological and Mining Engineering and Sciences Department at Michigan Technological University. His current teaching and research interests include hydrogeology, groundwater engineering, containment transport, and subsurface remediation. His approach to learning is to place importance on providing students opportunities in both the classroom and in research settings to develop their own problem-solving skills by attempting to solve problems independently. He is a member of the Association of Environmental Engineering and Science Professors, American Geophysical Union, National Ground Water Association, American Society of Civil Engineers, and Society of Mining, Metallurgy, and Exploration. He received his Ph.D. in environmental engineering from Michigan Technological University.

**Arthur Goldstein** is the founding dean of the College of Science and Mathematics at Bridgewater State University. Prior to joining Bridgewater State University, he held appointments as a dean at the University of New England and the director of the Division of Earth Sciences at the National Science Foundation (NSF). While at NSF, he was involved in developing GeoTeach, a program aimed at improving the development of preservice and inservice secondary school teachers. Prior to his appointment at NSF, he was a professor of geology at Colgate University and served as department chair for 5 years. He was the chair of the National Research Council (NRC) Committee on Trends and Opportunities in Federal Earth Science Education and Workforce Development and the cochair of an NRC study on scientific ocean drilling. He received a B.S. in geology from Kent State University and an M.S. and Ph.D. from the University of Massachusetts, Amherst.

**Edward Laine** is professor emeritus at Bowdoin College, interested in marine geology and oceanography. In Maine waters, his research topics include submarine groundwater discharge, hypoxia, and the record of Lake Wisconsin sea-level change. In the deep ocean, topics include the record of glacial erosion, radioactive waste disposal, terrigenous sedimentation, contourite drifts, and bottom current processes. Through the 2000 Problem-Based Service-Learning Institute sponsored by Maine Campus Compact, he saw a way to consistently and effectively form and nurture links between the campus and the community. He holds an A.B. from Wesleyan University and a Ph.D. from the Woods Hole Oceanographic Institute Joint Program in Oceanography at the Massachusetts Institute of Technology.

**Cathryn Manduca** *(committee chair)* is director of the Science Education Resource Center at Carleton College, where she is involved in a variety of projects that support improvements in undergraduate education and geoscience education. Her work includes organizing workshops and other activities for faculty and educators of all types. Topics of focus include bringing research results on teaching and learning into broader use in the geosciences, understanding geoscience expertise, and building strong geoscience departments. She is the executive director of the National Association of Geoscience Teachers and serves on the American Geophysical Union outreach committee, education committee of the American Institute of Physics, National Numeracy Network board of directors, and advisory board for gifted and talented education programs in her local school district. She is also the executive director of the National Association of Geoscience Teachers. She holds a B.A. in geology from Williams College and an M.A. and Ph.D. in geology from the California Institute of Technology.

**Suzanne O'Connell** *(presenter)* is a professor of earth and environmental sciences at Wesleyan University. She teaches geosciences with a strong emphasis on hands-on research with undergraduates. She uses research to encourage students from diverse backgrounds to consider careers in the geosciences. Her current research focuses on Antarctic climate change using sediment cores from the Weddell Sea Ocean Drilling Program Leg 113. Previously she spent a year at the Massachusetts Institute of Technology on a National Science Foundation Visiting Professorship for Women. She is a member of the American Geophysical Union, American Association for the Advancement of Science, Association for Women Geoscientists, and Geological Society of America. She received her Ph.D. on submarine depositional systems from Columbia University.

**Eric Riggs** *(presenter)* is associate dean and professor in the College of Geosciences at Texas A&M University. His research focuses on related aspects of teaching, learning, and cognition in the geosciences. His primary education research is on how to maximize the effectiveness of earth science education in critical areas for undergraduates, graduates, and professionals, and for underrepresented groups and communities with specific geoscientific needs. He previously served as president of the National Association of Geoscience Teachers and president of the American Geosciences Institute. He received his Ph.D. in geological sciences from the University of California at Riverside.

**John Saltmarsh** *(committee member)* is the director of the New England Resource Center for Higher Education (NERCHE) at the University of Massachusetts, Boston, as well as a faculty member in the Higher Education Doctoral Program in the Department of Leadership in Education in the College of Education and Human Development. He leads the project in which NERCHE serves as the administrative partner with the Carnegie Foundation for the Advancement of Teaching for Carnegie's elective Community Engagement Classification. He is an associate editor for the *Michigan Journal of Community Service-Learning* and serves on several editorial boards. He serves on the national advisory board of Imagining America and is a member of the board of the International Association for Research on Service-Learning and Community Engagement, National Review Board for the Scholarship of Engagement, and advisory committee for the Carnegie Foundation's Community Engagement Classification. In 2015, he was inducted into the Academy of Community Engagement Scholarship. From 1998 through 2005, he directed the national Project on Integrating Service with Academic Study at Campus Compact. He holds a Ph.D. in American history from Boston University.

**Suzanne Savanick Hansen** *(presenter)* is the sustainability manager at Macalester College and teaches in the Department of Environmental Studies. Her current research includes urban ecology, sustainability, and sustainability education. She previously developed the Sustainable Campus Initiative at the University of Minnesota and worked for the Science Education Resource Center and as a visiting professor at Carleton College. She worked for the Minnesota Office of Environmental Assistance and as an elected supervisor of the Dakota County Soil and Water Conservation District. She currently serves on the Forum of Women in the Environmental Fields' board of directors. She received her Ph.D. in conservation biology from the University of Minnesota.

**Linda Silka** *(presenter)* is senior fellow at the Senator George J. Mitchell Center for Sustainability Solutions at the University of Maine. She has experience in leading community-university research partnerships on environmental and environmental health issues, and she has written extensively on the challenges and opportunities of building research partnerships between communities and universities. She is a member of the Sustainability Solutions Initiative's Knowledge-to-Action Collaborative, which is investigating ways to overcome the barriers between knowledge and action and build partnerships between Maine's academic institutions and diverse stakeholders in order to better solve sustainability challenges. She is a member of the American Psychological Association and the Society for the Psychological Study of Social Issues. She received her Ph.D. in social psychology from the University of Kansas.

**Susan Sullivan** *(committee member)* is director of the education and outreach program of the Cooperative Institute for Research in Environmental

Sciences (CIRES) at the University of Colorado Boulder, which includes nearly 800 researchers and support people, including nearly 200 graduate and undergraduate students. She has helped develop programs to educate students of all ages and backgrounds about the geosciences and global and local environmental issues, including K-12 students, college undergraduates, students from underrepresented groups, and community college students. She has had ongoing engagement with the National Association of Geoscience Teachers through publications, presentations, and position statements. Her current projects include professional development workshops for science teachers, provision of education related to research projects, and oversight of numerous other education projects within the CIRES outreach group. Before beginning her work in K-12 education, she conducted research in atmospheric chemistry analytical methods with CIRES and the NOAA Aeronomy Laboratory. She holds a B.S. in chemistry from California Polytechnic State University, San Luis Obispo, and a Ph.D. in analytical chemistry from the University of Colorado Boulder.

**David Szymanski** *(presenter)* is associate professor of natural and applied sciences at Bentley University. He is a geologist with research interests ranging from the chemical evolution of magmas to environmental impacts of human and natural processes on surface waters. As a science communicator and educator, he conducts pedagogical research on how to teach nonscientists the business of applying scientific concepts to real-world problems. As a forensic chemist, he specializes in the examination of glass as trace evidence and has testified in a number of criminal cases. As a former congressional science fellow in the U.S. Senate, Szymanski now involves students in research for nonpartisan policy development on issues of energy, climate, and natural resources. He received his Ph.D. from Michigan State University.

# Appendix C

# Resources Suggested by Participants

| Resource | Information from Website | Link |
|---|---|---|
| National Science Foundation's Improving Undergraduate STEM Education: Education and Human Resources (IUSE:EHR).<br>• Potential NSF funding opportunity for service-learning and/or educational research on geoscience service-learning. | The IUSE:EHR program invites proposals that address immediate challenges and opportunities facing undergraduate STEM education and those that anticipate new structures and new functions of the undergraduate learning and teaching enterprise. | http://www.nsf.gov/funding/pgm_summ.jsp?pims_id=505082 |
| National Science Foundation's Improving Undergraduate STEM Education: Pathways into Geoscience (IUSE:GEOPATHS).<br>• Potential NSF funding opportunity for service-learning and/or educational research on geoscience service-learning. | The IUSE:GEOPATHS program invites proposals that specifically address the current needs and opportunities related to undergraduate education within the geosciences community. The primary goal of this funding opportunity is to increase the number of undergraduate students interested in pursuing undergraduate and/or post-graduate degrees in geoscience through the design and testing of novel approaches for engaging students in authentic, career-relevant experiences in geoscience. | https://www.nsf.gov/funding/pgm_summ.jsp?pims_id=505169 |
| National Science Foundation's Research Traineeship Program (NRT-IGE track)<br>• Potential NSF funding opportunity for service-learning and/or educational research on geoscience service-learning. | The IGE track of the NRT focuses on test-bed projects aimed at piloting, testing, and validating innovative and potentially transformative approaches to graduate education. This track supports testing of novel models or activities with high potential to enrich and extend the knowledge base on effective graduate education approaches. | http://www.nsf.gov/funding/pgm_summ.jsp?pims_id=505015 |

| | | |
|---|---|---|
| InTeGrate Program (Interdisciplinary Teaching about Earth for a Sustainable Future) | This workshop program provides materials on approaches to measuring service-learning outcomes. It includes links to the Community Service Attitudes Scale, the Inquiry & Analysis VALUE Rubric, and the RISC Survey. | http://serc.carleton.edu/integrate/programs/implementation/program3/workshop2/program.html |
| Association of American Colleges & Universities (AAC&U) VALUE Rubric Development Project | As part of AAC&U's LEAP initiative, the VALUE rubrics contribute to the national dialogue on assessment of college student learning. Each rubric was developed from the most frequently identified characteristics of learning. | http://www.aacu.org/value/rubrics |
| Science Education for New Civic Engagements and Responsibilities (SENCER) | SENCER applies the science of learning to the learning of science, all to expand civic capacity. Their courses and programs connect STEM content to critical local, national, and global challenges. They improve education by focusing on real-world problems and extend the impact of this learning to the broader community and society. | http://www.sencer.net/ |
| Michigan Technological University D80 | The D80 programs exist to work on solutions for prosperity with partnering communities around the world. The website provides a virtual center with links to different development-related service learning projects. | http://www.mtu.edu/d80/ |

continued

| Resource | Information from Website | Link |
|---|---|---|
| Michigan Technological University Student Learning Goals | Service-learning relevant learn-outcomes/assessment rubrics | http://www.mtu.edu/assessment/program/university-learning-goals/goal-8-rubric.pdf<br>http://www.mtu.edu/assessment/program/university-learning-goals/ |
| Laboratory for Civic Engagement Volunteering, Service Learning, Civic Engagement, Public Scholarship – What does it all mean? | Workshop at Pennsylvania State University-Brandywine where the campus has defined the differences in the "jargon" between volunteering, service learning, civic engagement, and public scholarship. | http://sites.psu.edu/wp-content/uploads/sites/14328/2012/01/civicengagementdefinitions.pdf |
| Macalester College's Civic Engagement Center | The Civic Engagement Center educates students on issues related to global citizenship, assisting them with their personal discernment of values, ethics, and commitments towards contributing to the public good in meaningful ways. | http://www.macalester.edu/cec/ |
| Stanford graduate student teaching example | The primary objective of the InTeGrate program at Stanford is to recruit and increase the retention of underrepresented minorities within STEM. The objectives are worked towards by exposing undergraduate students in early geoscience courses to address global sustainability issues, and creating URM postdoctoral advanced graduate students to teach these modules in classrooms. | http://serc.carleton.edu/integrate/programs/implementation/program6/index.html |

| | | |
|---|---|---|
| American Geophysical Union's Thriving Earth Exchange | Thriving Earth Exchange helps communities leverage Earth and space science to build a better future for themselves and the planet. This program links communities to science expertise. | http://thrivingearthexchange.org/ |
| Indiana University-Purdue University at Indianapolis, Center for Service and Learning | The Center for Service and Learning builds partnerships between students and communities within Indianapolis. Institutional-wide example of Centers that facilitate service learning. | http://csl.iupui.edu/ |
| Indiana University-Purdue University at Indianapolis, Center for Earth and Environmental Science (CEES) | The CEES Environmental Service Learning Program focuses on raising awareness of ecological health issues as they relate to water quality. The projects provide students with opportunities to directly experience many topics discussed in courses as well as observing how communities work together to solve environmental issues. | http://cees.iupui.edu/environmental-service-learning-undergrad |
| Bentley University's Service-Learning and Civic Engagement Center | The Bentley Service-Learning center enhances student learning through academically relevant service and work with local, regional, and international nonprofit community to address important public issues. | http://www.bentley.edu/centers/service-learning-center |

*continued*

| Resource | Information from Website | Link |
|---|---|---|
| A History of Service-Learning | University of Southern California College of Letters, Arts and Sciences article describes how service-learning became pedagogy and provides links to what they consider the fundamental principles to help faculty and others understand the concepts on a general level. | https://dornsife.usc.edu/history-of-service-learning/ |
| Earth Science Week | Affiliated with the American Geosciences Institute, Earth Science Week is a national and international event that helps the public gain a better understanding and appreciation for the Earth Sciences and to encourage stewardship of the Earth. Links provided to see global participation opportunities. | http://www.earthsciweek.org/ |
| Geoscientists Without Borders (GWB) | Geoscientists Without Borders supports humanitarian applications of geoscience around the world. GWB places geoscientists around the world where they can mitigate the results of natural hazards in the neediest communities. | http://seg.org/About-SEG/Geoscientists-Without-Borders |

NOTE: All above links were accessible as of January 2017.